Prairie Dog Empire

Other Titles by Paul A. Johnsgard
Published by the University of Nebraska Press

Prairie Dog Empire

A Saga of the Shortgrass Prairie

PAUL A. JOHNSGARD

University of Nebraska Press, Lincoln and London

Library of Congress Cata-
loging-in-Publication
Data
Johnsgard, Paul A.
Prairie dog empire : a saga
of the shortgrass prairie /
Paul A. Johnsgard.
p. cm.
Includes bibliographical
references and index.
ISBN 0-8032-2604-7
(hardcover : alk. paper)
1. Prairie ecology – Great
Plains. 2. Prairie animals –
Great Plains. I. Title.
QH104.5.G73J64 2005
578.74'4'0978 – dc22
2004024696
Set in Minion by Kim
Essman.
Designed by Dika
Eckersley.
Printed by Thomson-
Shore, Inc.

Dedicated to the people and groups who have long and valiantly worked to preserve our native grassland heritage

Contents

TABLES

Preface

In early July of 2002 I attended a hearing at the Lincoln headquarters of Nebraska's Game and Parks Commission. The main agenda item at this hearing was whether the State of Nebraska would begin to take steps toward conserving the black-tailed prairie dog. As a result of a 1998 petition from the National Wildlife Federation, followed up later by one from the Biodiversity Legal Foundation, the species had been legally proposed as a candidate for listing as nationally threatened. Nebraska is one of eleven states within the species' current range and had been previously notified by the U.S. Fish and Wildlife Service that such a listing, with all its associated required conservation measures, could be averted only if these states individually and collectively began concerted efforts to begin preserving the species.

An hour or more of oral testimony by conservationists, biologists, and representatives of various environmental groups ensued, which uniformly favored initiating a modest conservation program, including some supportive testimony from biologists of the Game and Parks Commission itself. Then a group of ranchers who had been bused in from across the state took the floor, condemning the prairie dog with all the usual vituperation that has been associated with rancher–prairie dog relationships of the past century or more. At the end of the meeting the commissioners unanimously voted not to provide any conservation efforts whatsoever toward prairie dogs and furthermore voted to terminate all state-supported research on the species' status and biology that was then being undertaken or planned by the commission's staff biologists. The ban on future prairie dog studies was later rescinded, but no steps toward producing a conservation management plan were taken.

As a result of that meeting, I decided that enough interest exists in prairie dogs, including strongly held attitudes both pro and con, to warrant a book centered on this controversial animal. Such a book would also describe and document the semiarid grassland ecosystem of the western plains within which the prairie dog evolved and in which it has historically played a pivotal, or "keystone," ecological role. In many ways, by the late 1900s the black-tailed prairie dog was providing a reprise of the sad history of the North American bison, whose genocidal destruction during the late 1800s spelled the end of the American frontier and likewise brought an end of the entire culture that had been represented by the bison-dependent Dakotas and other Native Americans of the western plains. Now, roughly a century later, many ranchers who had displaced the Native Americans were themselves being threatened, as a combined result of prolonged droughts, long-depressed cattle markets, and ranges so badly overgrazed and weed infested that most of the small operators could not provide the numbers of livestock needed to make a living.

This book is a result of these experiences and related concerns. As an ornithologist I have become increasingly worried about the decline and often regional disappearance of the burrowing owl, a commensal associate of the prairie dog. I also fear for the ferruginous hawk, whose nest sites and winter distributions in the Great Plains and American Southwest often center on prairie dog "towns," and for the tiny swift fox, an agile and beautiful little carnivore that once also was common but now is largely confined to areas of prairie dog towns. The swift fox is presently so rare (as of 2004 it is a federal candidate species for a listing as threatened) that I have yet to see one in the wild. By the time I began my writing, it was already perhaps too late to save the black-footed ferret, another prairie dog–dependent carnivore that is already federally listed as endangered and that is surviving only by the thinnest of life-support threads. The ferret, one of America's rarest mammals, was one of the first mammal species to be placed on the nationally endangered list. Nevertheless, the North American bison was also once reduced to perhaps no more than a hundred or so animals surviving in the wild. Yet the species has rebounded and now numbers several hundred thousand animals, mostly in captive herds.

All of these symbols of the western plains have close ecological ties to prairie dogs. If prairie dogs are not to be saved, we have little hope of preserving some of the others either, to say nothing of affecting the dozens of additional animals that are also part of and variously dependent on the prairie dog–bison–buffalo grass ecosystem of the American plains.

I have relied on the help of many people in assembling the information needed for this book. Among them are Craig Knowles of FaunaWest Wildlife Consultants, who provided me with several valuable unpublished reports on the status of three species of prairie dogs. Robert Luce, interstate coordinator, Prairie Dog Conservation Team, also provided me with some important references and an advance copy of the proceedings of a Conservation Team workshop meeting held in late 2003. Tyler Sutton of the Great Plains Conservation Alliance sent me various unpublished reports on prairie dogs and plans for establishing a high plains ecological preserve south of the Black Hills. "Buffalo" Bruce MacIntosh and Gerald Jasmer provided me with valuable unpublished materials on the prairie dog and the swift fox. Al Steuter of the Nature Conservancy's Niobrara Valley Preserve loaned me materials on bison biology and bison rearing. Joe Truett of the Turner Biodiversity Fund also provided me with unpublished information on the foundation's important conservation and restoration programs, and Francis Moul gave me a manuscript copy of some of his graduate research on the history of the national grasslands. Barbara Voeltz of the Nebraska Game and Parks Commission helped me with Internet searches of the extensive biological literature of the western plains grasslands. I also wish to thank the three anonymous persons who reviewed the manuscript for the University of Nebraska Press.

I decided that before I finished writing I needed to visit some of the remaining natural grassland regions within the vast Great Plains range of the black-tailed prairie dog, and especially the major National Grassland reserves that are at the heart of the species' remaining range. For advice or help with fieldwork, providing reference materials, and other assistance, I am indebted to Doug Backlund, Linda Brown, Jackie Canterbury, Bob Gress, Josef Kren, Bob Luce, Al Steuter, Tom Shane, and Scott Wendt. All of these people helped me understand the often neglected and invariably sad histories of all of the historic residents of our western American prairies, from prairie dogs, wolves, and bison to our Native American kin, whose lives and fortunes were fatally and inexorably intertwined with the prairie.

In a broader sense I owe a special debt of gratitude to the organizations that are fighting to preserve the prairie dog and its associated prairie ecosystem as part of our American heritage and that have helped to document their status. These groups include such diverse public interest organizations as Biodiversity Associates, the Biodiversity Legal Foundation, the Center for Native Ecosystems, the Conservation Alliance for the Great Plains, Forest Guardians, the Fund for Animals, the Humane Society, the National Wildlife Federation, The Nature Conservancy, the Prairie Dog Coalition, the Predator Conservation Alliance, Rocky Mountain Animal Defense, and the Turner Biodiversity and Endangered Species Funds. Without the documented information and related historic materials these groups have assembled, and without their legal maneuvering against entrenched federal agencies and powerful opposing private interests, this book might not have been written. Without their efforts, not enough of the prairie dog ecosystem would be left to bother to preserve, and the story of the prairie dog would have to be told only in the past tense.

Prairie Dog Empire

The Western Shortgrass Prairie

A Brief History

There was nothing but land. Not a country at all, but the materials out of which countries are made. – *Willa Cather,* My Antonia

---

Like the animals and plants that now live on it, the land composing the surface and substrate of the North American Great Plains nearly all came from someplace else. Some of the region's parts were carried by westerly winds as volcanic ash from mountains a thousand or more miles to the west, while others were blown in as dust-sized particles from areas up to several hundred miles to the northwest. Most of what geographers call the Great Plains was carried in and deposited by rivers and streams originating in the Rocky Mountains or was randomly strewn as glacial-carried materials from the north. Only in a few areas, such as the igneous and metamorphic Precambrian core of South Dakota's Black Hills and the 300 million-year-old (Paleozoic) Wichita Mountains of Oklahoma, are the remnants of the earth's early violent history exposed in the form of ancient and eroded mountains (Trimble 1990).

In its simplest form, the present-day Great Plains region may be thought of as a poorly constructed tabletop that is slightly tilted downward toward the east-southeast. Its western limits are formed by the towering Front Range of the Rocky Mountains, cresting at the Continental Divide, and grading into their outlying piedmonts, which end indecisively at about 5,000 to 6,000 feet of elevation. From there the Great Plains take over, gradually losing altitude eastwardly. At roughly 1,000 feet of elevation, the Great Plains merge invisibly with the Central Lowlands, a region massively shaped by glaciers at the northeastern end and forming a shallow basin made up of the Missouri, Mississippi, and Ohio river valleys toward the southeast. These river valleys were cut and repeatedly reformed by the immense meltwaters of several glaciations, carving broad, fertile valleys as they slowly traveled to deposit their silty cargoes along the way or dump them into the Gulf of Mexico.

The southern parts of the Great Plains become increasingly flat and arid, including the treeless and so-called Staked Plains of the Texas Panhandle and eastern New Mexico, a land so flat and barren that early explorers reputedly drove tall wooden stakes into it at intervals to keep from becoming lost. The southern terminus of the Great Plains is decisively marked in New Mexico by the Mescalero Escarpment and the associated Pecos Valley, and in Texas the Plains are also abruptly terminated southwardly by the Edwards Plateau. The northern end of the Great Plains disappears quietly somewhere near Great Slave Lake in Canada's vast transcontinental coniferous forest, an enormous area largely scraped down by glacial action nearly to ancient bedrock, the Precambrian Shield of the earth's earliest crust.

Within the Great Plains region, the higher, more arid western portions that now support only shortgrass and mixed-grass prairies dominated by perennial herbs of low to middle stature can be geologically and biologically distinguished as the High Plains. Geologically speaking, the northern edge of the High Plains is the Pine Ridge Escarpment of northwestern Nebraska and southwestern South Dakota. From there south, most of the High Plains consists of a nearly flat surface, covered by sandy materials called the Ogallala formation, a layer up to several hundred feet thick of water-transported materials that were deposited in Pliocene times. The upper part of this layer is commonly a thick zone of very hard carbonates, producing a "caprock" of rocklike material (caliche) that is almost impervious to water. This zone, not very evident in the northern part of western Nebraska's western plains, becomes thicker southwardly and is up to 30 feet thick at the southern end of the western plains in Texas. The conspicuous Caprock Escarpment, extending in a general north-northeast to south-southwest direction from western Oklahoma through the Texas Panhandle, marks the southeastern edge of the caprock zone and provides a convenient southern bookend to the High Plains (Thornbury 1965).

North of the Pine Ridge Escarpment and the nearby Black Hills is the Missouri Plateau of eastern Montana, northeastern Wyoming, and the western Dakotas, and the Missouri River valley itself. Eastward and northward from the Missouri River in the Dakotas is a fertile region of glacial drift and till deposited by the last glaciation, forming a soil substrate for the biologically rich prairie wetlands region of the northern plains. Although technically not a part of the High Plains as they have been geologically defined, these portions of the western and central Dakotas, eastern Montana and Wyoming, and the southernmost prairie-covered portions of Alberta and Saskatchewan share most of the same semiarid and grassland-adapted plant and animal life that is now found in the more southern parts of the Great Plains, including a keystone species, the black-tailed prairie dog. For this book's purposes these areas are included in the general region here informally recognized as the western plains, or "shortgrass prairie ecosystem."

To understand the current overall patterns of plant and animal distributions in the western plains, one must think backward a few million years. Until about 70 million years ago, all of what is now the Great Plains lay beneath a vast inland sea. At its bottom were sediments up to 10,000 feet deep. Around 70 million years ago the North American continent began a gradual uplift, progressively exposing the sea bottom and producing a series of gentle basins and arches on the rising landscape. However, more extensive uplifting occurred in the Black Hills region, and the Rocky Mountains to the west began their much more extensive expansions, largely through folding and faulting. As rapidly as the western mountains rose, rivers formed to erode their surface layers, carrying sediments away to the more easterly vegetated plains, where a variety of plant-eating and carnivorous dinosaurs were dining on what would – by about 65 million years ago – became their final meals. The river sediments preserved the fossil remains of such impressive but now defunct animals as *Triceratops* and *Tyrannosaurus*.

About 50 million years ago volcanic activities in mountains far to the west added additional wind-carried materials, even as abundant tropical vegetation still flourished along the edges of the retreating seas. In places like the northern plains these lush forests of tropical evergreen vegetation were eventually buried to become the vast beds of lignite and coal that are being extracted today from Wyoming, Montana, and the western Dakotas.

Between about 45 million and 35 million years ago a pause in mountain-building occurred. Gradually drier and cooler climates and less-forestlike plant communities began to develop, only to be interrupted by a renewal of volcanism and mountain building near the end of this period. More river-related erosion brought new layers of sediments to the Plains, including the White River sediments that compose much of South Dakota's scenic badlands. By now the land east of the Rocky Mountains was mostly covered by a vast grassland and many large grazing mammals. Early horses and camels had displaced the dinosaurs and other dominant reptiles of the late Mesozoic era. Fossil remnants of some of these high-toothed grazing mammals were preserved in such places as Agate Fossil Beds National Monument in western Nebraska, among sediments more than 20 million years old.

Between 25 and 5 million years ago the Great Plains were affected by periodic phases of mountain building and volcanism, each episode adding layers of new sediments and windblown debris to the older materials progressively interred below. The oldest deposits of sediments occurred in the northern parts of the Great Plains, while more recent deposits of Oligocene, Miocene, and Pliocene vintages extended progressively farther east and south, eventually reaching the southern edge of the present-day Great Plains. A rather late surge of volcanism in the western mountains occurred about 9 million years ago. This activity suffocated herds of rhinos and other large mammals under a cloud of volcanic dust, providing the fossilized base for what is now known as Ashfall Fossil Beds State Historical Park, in north-central Nebraska. By 5 million years ago the entire High Plains region was becoming a vast, eastwardly sloping tableland similar to its present shape, with numerous rivers – including the predecessors of the Missouri, the Platte, the Cimarron, and the Pecos – all cutting downward through the soft surface materials near the mountains and redepositing them farther downstream.

About 5 million years ago a new period of uplift occurred in western North America, causing new cycles of down-cutting and subsequent deposition, with rivers such as the Missouri eroding down to depths only a few hundred feet higher than their current elevations. Prior to the Pleistocene glaciations, the Missouri River had flowed northeastward into Hudson Bay, probably cutting across northern Montana and entering Canada in southeastern Saskatchewan. Its northern tributaries, including the Yellowstone, the Little Missouri, and the now isolated Souris, were part of this same general river system, as perhaps also were the Knife and the Cannonball rivers. Then, about 2 million years ago, the first of a series of continent-wide glaciations developed across most of northern North America, with roughly the northeastern fifth of the Great Plains becoming repeatedly covered with sheets of ice hundreds of feet thick. At maximum, the glaciers' southernmost limits in the Great Plains extended from the general vicinity of the Rocky Mountain foothills in present-day northwestern Montana eastward and southward along a line approximating the present southeastward-flowing course of the Missouri River, to eastern Nebraska and northeastern Kansas.

With the glaciers came a host of cold-adapted mammals, including mastodons, mammoths, bison, caribou, muskoxen, wolves, tiger-sized cats, bears, and many other cold-adapted species (Schultz 1934). Along the glaciers' southwestern edges in what is now the Dakotas, ridges of rock debris produced terminal moraines that are collectively called the Coteau du Missouri. Irregularities left by stagnant ice produced pockmarked terrains in northeastern Montana and the Dakotas that eventually would become the famous "duck factory" or "prairie pothole" region. The Missouri River began to form its present, generally southward flowing course, and rivers like the Platte and Niobrara also shifted their courses as they were variously influenced by glacial moraines and meltwater effects. The current channel of the Missouri closely follows the southwestern limits of the last (Wisconsinian) glaciation. Highly sculptured badlands in the western Dakotas were also produced, mostly by water erosion from glacier-fed rivers that now sometimes are barely more than seasonal trickles.

The plant and animal life present during this broad swatch of North American history was probably an ever-changing tapestry, as climate patterns ebbed and flowed and as the land itself was reshaped. It is clear that the general pattern of climatic change in the Great Plains, as elsewhere in North America, has been one of general cooling and drying, at least since the end of the Age of Dinosaurs (the Cretaceous era) some 65 million years ago. As mentioned, the tropical forests that once covered the great interior lowlands of North America disappeared and were very gradually replaced by more drought- and cold-tolerant forms. Thus tropical evergreen trees were replaced by seasonally deciduous trees, deciduous trees by shrubs, and shrubs by herbs (broad-leaved herbaceous plants and grasses). Grasses, the plant type best adapted to short growing seasons, recurrent fires, and grazing by large animals, were destined to become the dominant plant type of the Plains. Somewhere between the middle Miocene epoch (about 15 million years ago) and historic times, the Great Plains were converted from a semiopen forest, with only scattered grassy areas, to a completely open grassland, with forests and woodlands limited to watercourses and other areas protected from fire and having access to surface or subsurface moisture (Axelrod 1985).

Plants living in northwestern Mexico, a region of long-term deserts and semideserts associated with the permanent high-pressure zones of these latitudes, probably provided a reservoir of plants and animals pre-adapted to an arid-zone pattern of existence. From this general region, especially the present-day Chihuahuan Desert, various grasses and other ecologically similar plants presumably moved north to occupy the increasingly significant rain-shadow effect of the Rocky Mountains, where most of the clouds carrying moisture from the Pacific Ocean were nearly squeezed dry as they passed eastward over the high western cordilleras.

The lowly but pervasive buffalo grass, which now extends from the southern tip of the Chihuahuan Desert north across the drier plains almost to the Canadian border, is one of the many arid-adapted grasses that probably had its origins here, and it has certainly been a foraging mainstay for such grazing mammals as bison and prairie dogs for uncountable millennia. The Mexican prairie dog is a southern representative of the keystone prairie dog assemblage of five grassland-dependent species and likewise is a part of this same longstanding semidesert community. It also is perhaps a Mexican relict of an original arid grassland distribution pattern of ancestral prairie dogs. It still survives in a southern corner of the Chihuahuan Desert, although it is now an endangered species. The coyote may have originated in the Mexican plateau too; in the

Aztec culture, Heuheucoyotl was a trickster deity and Coyolxauhqui was the moon goddess. Our English name for the coyote is thus derived from the Aztec *coyotl*, a fitting tribute both to its cunning nature and to its love for singing on moonlit nights. Likewise the kit fox, a close kin to the more northerly swift fox, has a semitropical desert grassland orientation, as do several jackrabbit species, such as the white-sided jackrabbit. The adjacent dry mountain slopes of the Sierra Madre Occidental and Oriental, where scrub oaks, mesquite, and similar arid-adapted woody plants still occur, probably provided various shrubs and deciduous trees to the mix of plant species that were slowly inching their way northward over the western plains.

The climate of the western plains gradually settled into a pattern character-ized by dry, cold winters, with much of the cold-season precipitation falling as snow. Most of the annual precipitation occurred during the spring and sum-mer growing season, often in the form of violent thunderstorms that occurred as moisture-laden warm air from the Gulf of Mexico met cooler and drier air currents coming southeast out of the northern interior. Increasingly short sum-mers occurred northwardly, and increasing amounts of annual sunshine and associated evaporation occurred toward the south. However, periodic droughts and recurrent, lightning-caused prairie fires added to the complexities of life for organisms adapting to these demanding conditions for survival. The grasses and other plants of the developing shortgrass and mixed-grass prairies were ones that, like the desert and desert grassland plants of the Southwest, were mainly warm-season perennial species, whose photosynthetic processes oper-ate well under conditions of high summer temperatures and extended periods of relatively little moisture, sometimes lasting for several years. Among these species are such important southern and western shortgrass and mixed-grass prairie genera as *Andropogon, Buchloë, Bouteloua, Chloris, Paspalum, Setaria, Schizachyrium, Sorghastrum,* and *Sporobolus.* Plants that evolved to form the tallgrass prairies at the northeastern, moister edges of the Great Plains have a higher proportion of cool-season grass genera. These include such typical prairie grasses as *Agropyron, Bromus, Elymus, Festuca, Hordeum, Poa,* and *Stipa* (Sims 1988). At the northernmost edge of the Great Plains, *Festuca* grasses became especially important in the transition zone toward the forests of the Rockies and the Canadian boreal forests. These tall, productive grasses are well adapted to cooler summer temperatures and greater precipitation than are most mixed-grass prairie grasses. As a result, bison of the Canadian plains histori-cally moved west toward the Front Range of the Rocky Mountains, rather than migrating great distances southward, to obtain their fall and winter sustenance from this nutritious forage source.

The present climate of the western plains may be summarized relative to its geographically variable precipitation. At the western edge of the plains an annual precipitation of less than 16 inches is typical. The mixed-grass prairies merge eastwardly with the tallgrass prairies very roughly at the isohyet of about 28 inches of annual precipitation.

Various ways exist to represent the geography, climate, and associated biological communities of the Great Plains visually. One current and widely adopted method was first proposed by Robert Bailey. It recognized twenty-five separate ecological-geographic provinces in the United States south of Canada. Bailey's boundaries were refined by The Nature Conservancy (1999) into fifty-two ecoregions within the same forty-eight-state boundaries, of which five fall within the previously designated geographic limits of the western plains (Map 1). Of these five the Black Hills are not geologically a part of this region, leaving the Northern Great Plains Steppe, the Central Mixed-grass Prairie, the Central Shortgrass Prairie, and the Southern Shortgrass Prairie as constituent parts of the western plains. Some of the limits of these ecoregions have since undergone further minor refining, but the collective boundaries of these five ecoregions make for a convenient visualization of the entire western plains biological domain and will be used later in the book as a collective biogeographical definition of this region.

More recently a group of avian biologists formed the North American Bird Conservation Initiative Committee (NABCIC), which operates through regionally based, biologically driven, and landscape-oriented partnerships. A team from that group assembled a map of proposed Bird Conservation Regions (North American Bird Conservation Initiative Committee 2000). This map divides the Northern Great Plains Steppe into a more northerly "Prairie Potholes" component, which extends north into central Alberta and east into the tallgrass prairies of the Dakotas, Minnesota, and Iowa, and a more southerly and drier "Badlands and Prairies" segment. The other western plains elements recognized in the map are the southwestern Shortgrass Prairie region and the southeastern Central Mixed-grass Prairie segment. Three of these Bird Conservation Regions are mapped here (Map 2); the map excludes the Prairie Potholes segment, as it mostly falls outside the recognized geographic limits of the western plains.

The last map reproduced here includes the entire Great Plains and part of the Central Lowlands of North America (Map 3). It is largely based on a National Wildlife Federation historical vegetation-based map (Bachard 2001), which is geographically limited to the native perennial grasslands (tallgrass, mixed-grass, and shortgrass prairies) of interior North America. Additionally, the Nebraska

Sandhills are designated separately in Map 3; they are botanically and zoologically quite recognizable, and their separate inclusion provides a convenient geographic separation between the biologically distinguishable northern and southern segments of the mixed-grass prairies.

It is one thing to look at a map and try to visualize the kinds of landscapes represented therein, to say nothing of the diversity of plants and animals that might be implied in a simple map label such as "shortgrass prairie." For example, the native prairies of Buffalo Gap National Grassland in the southwestern South Dakota badlands region are known to support 230 species of birds, 58 mammals, 19 reptiles, and 8 amphibians as well as 92 grasses, 160 broad-leaved herbs (or forbs), 22 shrubs, and 11 trees. Beyond these are uncountable numbers of insects, other invertebrates, and soil microorganisms. Most of this very land was once ranched or farmed for only a few decades, from when it was first settled in the later 1800s until the Dust Bowl years of the 1930s, when bankruptcies became as common as summer dust storms. Plowing up such fragile lands to raise wheat or corn for a few decades, often until the topsoil blows away and the land is abandoned, is like throwing away a treasure trove of potential biological riches to raise a single species of grass that needs so much tilling, water, herbicides, and pesticides to survive that scarcely anything else of value can survive there.

One's native land is the most important thing on earth. Above all it is made holy by the ancestors, who pass it on. – *Louis Tiel*

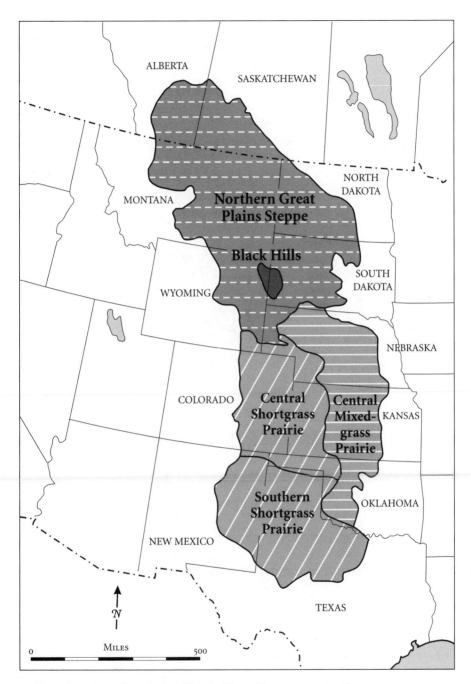

Ecoregions of the Great Plains. After the Nature Conservancy (1999).

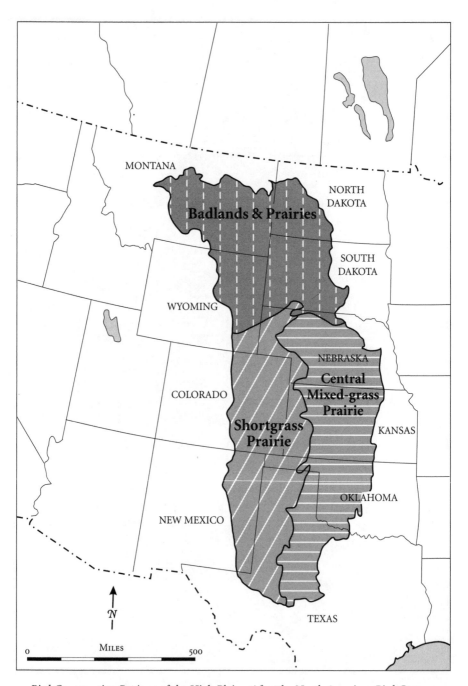

Bird Conservation Regions of the High Plains. After the North American Bird Conservation Initiative Committee (2000).

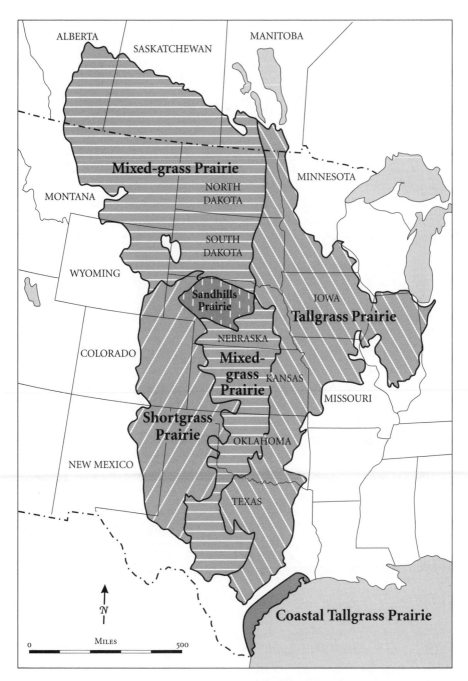

Biological regions of the Great Plains. Mainly after Bachand (2001) but with the Sandhills Prairie additionally designated.

2

A Buffalo Nation

Its Death and Rebirth

I ascended to the high country and from an eminance I had a view of a greater number of buffalo than I had ever seen before at one time. I must have seen near 20,000 of those animals feeding on the plain. – *Captain William Clark, August 29, 1806, writing in what is now central South Dakota*

There is such a quantity of them that I do not know what to compare them with, except the fishes of the sea. . . . We could not see their limit either east or west. . . . The plains were black, and appeared as if in motion. . . . The country was one robe. – *Various historical notes collected by Joel Allen (1876)*

Words fail to describe the uncountable numbers of bison that roamed the North American plains during the early 1800s. Estimates of 30 million to 40 million are most common, a number at least fifty times greater than the number of wildebeests found on the high savannas of eastern Africa in recent years. According to Tom McHugh, one single herd that was seen in 1862 by Nathaniel Langford while crossing what is now North Dakota was estimated to be 5 to 6 miles wide and required an hour to pass Langford's party at an average rate of about 12 miles per hour. This herd was estimated to contain a million animals. Another herd, seen along the Santa Fe Trail by Thomas Farnham in 1839, appeared to be some 30 miles wide and took three days to pass while Farnham was traveling through the multitude at a rate of about 15 miles per day. The number of animals that must have been in that single herd is simply beyond comprehension.

Ernest Thompson Seton (1909) once estimated that about 3 million square miles of the plains and prairies were occupied by bison, with at least 40 million on the High Plains, 30 million on the prairies, and about 5 million elsewhere. This total of 75 million animals now seems extravagant to most observers. Henry Epp and Ian Dyck (2002) have recently suggested that the total may have been more like 30 million animals, including about 24 million that were seasonal migrants plus 6 million that were essentially residential. These herds were utilized by a one-time Native American population estimated by Epp and

Dyck at some 86,000 to 130,000 hunters. A collection of bones representing more than six hundred bison – the Hudson–Meng Bison Bonebed – has been found in northwestern Nebraska. It is thought to be about 9,500 years old, perhaps representing a single large bison kill or several smaller ones that occurred within a short time period. This was probably the site of a buffalo jump, in which bison were driven over a cliff into a ravine, where the injured animals could then be killed. Such buffalo jumps were often used in suitable sites on the High Plains for killing large numbers of animals in a short time. Nearby are the fossil remains of two Columbian mammoths from about the same time period, their tusks locked together in mortal combat.

Frank Roe (1970) suggested an early historic bison population of about 40 million animals, whereas Tom McHugh (1972) thought there could have been no more than 30 million, inasmuch as an average density of 25 acres of grassland would be needed to support a single bison (10 acres in the tallgrass prairie, 45 acres in the shortgrass plains). An estimated total historic grassland area of 1.25 million square miles was used for McHugh's computations, and an equivalent reduction of 4 million animals was used to account for competition from coexisting deer, elk, and pronghorn. Andrew Isenberg (2000) has likewise estimated the historic Great Plains grassland ecosystem at about 1.2 million square miles (or 3.1 million square kilometers), and P. L. Sims (1988) came up with a similar figure (3.4 million square kilometers). Like the bison and other native plains megafauna, most of this area has now vanished, including about 95 percent of the tallgrass prairies as well as roughly 75 percent of the combined mixed-grass and shortgrass prairies. Most of the remaining prairies of the Great Plains consist of shortgrass, but even these have largely been degraded beyond easy salvation by cattle overgrazing.

Bison were the very icon of the American West, and the cultures and very survival of many tribes of Native Americans were intimately interconnected with them. For the Oglala Dakota the bison was Tatanka, the chief of all animals, and represented the earth itself. The awesome physical power of the bull was self-evident and was imitated by men in their use of buffalo robes and bison headdresses during special rites of the Buffalo society. The impressive sexual power of the huge bison males over the females was also not lost on Oglala males, nor was the willingness of a male to fight furiously with other males when needed to achieve and maintain access to his females. The maternal care of the female bison similarly offered lessons for women in caring for their own young, and the role of female bison in leading moving herds also set an example. Many Plains tribes regarded the bison as an integral part of their Creation myths, at times helping to explain the creation of humans as well as providing for their perpetual well-being and sustenance.

Out of the hole in the ground came the buffalo. Bulls and cows without number. They spread wide and blackened the plains. Everywhere I looked great herds of buffalo were going in every direction, and still others without number were pouring out of the hole in the ground to travel on the wide plain. – *Crow Chief Plenty-coups, as told to Frank B. Linderman (1962)*

This endless bounty of bison could not last forever, especially after the West began to be settled and as the effectiveness of rifles, ammunition, and gunpowder improved and guns became more widely available following the Civil War. Francis Haines (1995) has estimated that it may have required seven to ten bison annually to feed each Native American of the bison-hunting tribes, and that 2 million animals per year would have sustained them all. A herd of 30 million bison should produce about 6 million calves per year, or more than enough to cover the herd's losses to Native hunters. Wolves probably represented the bison's only other significant predator prior to the arrival of European American hunters.

Early colonists of the eastern coastal region, such as those from Virginia, probably began eliminating bison as best they could to provide more grazing lands for their cattle and perhaps to produce an occasional source of food. But flintlock-era guns of the day were a poor match for bison, and it is likely that the retractions of the bison's range east of the Mississippi between 1600 and 1800 was at least in part a voluntary retreat westward on the part of the bison into the heart of their preferred range, the great interior grasslands. Yet by the late 1700s a trade in bison hides was developing, as early explorers brought back hides tanned by and obtained from Native Americans. The luxurious hides were considered perfect as warm blankets for sleighs, carriages, and beds (Haines 1995).

Between about 1810 and 1860, as the railroads progressively pushed westward toward the Mississippi and Missouri rivers, the range of the bison shrank dramatically to about half its former size (Map 4), and the numbers of animals also may have substantially diminished, although the peak in hunting activities did not occur until after the Civil War. Two bison killed in central Wisconsin in 1832 may have been the last remnants of the herds east of the Mississippi River. By then, western-bound travelers and hunters were making inroads on the Great Plains herds, and the Native Americans themselves were participating in the hunts to obtain more of the valuable buffalo hides. James Shaw (2000) believed that bison living west of the Mississippi still numbered in the tens of millions at the close of the Civil War but that no more definite number could be assigned for them. By the 1870s single-shot breech-loading cartridge rifles had largely replaced their muzzle-loading predecessors, and .44- or .45-caliber ammunition became the weapon of choice for the commercial bison hunters. Only by 1875 did Buffalo Bill's favorite .50-caliber ammunition become available, and by then the bison herds had been virtually eliminated (McHugh 1972).

By 1871 innovations in the tanning process used to treat dried bison hides sparked a new interest in bison leather for uses in machinery, clothing, furniture, and vehicle cushions and tops. This booming market sent ever-greater numbers of hunters out into the Plains, now more easily accessible owing to railroad construction. One Kansas hunter killed three thousand bison in a single season in 1872, another hunter averaged a hundred animals a day for a month in 1872, and three brothers killed seven thousand bison during four months the same year. Some two hundred thousand skins were shipped out of a warehouse in Dodge City from 1872 to 1873, the first years that railroad freight east was available from that frontier town. About two hundred carloads of bison hindquarters were also sent east from Kansas that same year, as were two carloads of cured bison tongues (McHugh 1972).

Farther south the Texas herds were being dispatched with equal grim efficiency. One Texas hunter with his crew of four returned to Fort Griffin in the spring of 1878 with 4,900 bison hides. However, during the following winter he obtained only 800 hides. Other hunters in the same region also reported declining takes, and by 1879 most of the hunters had abandoned Fort Griffin. The inroads of the Kansas Pacific and the Santa Fe had already done their part in exposing the southern underbelly of the Great Plains' bison herds to the seemingly endless hordes of hunters. Many hunters probably moved north to exploit the herds still remaining north of the Platte River, where construction of the Union Pacific Railroad, after having reached Iowa in the 1850s and moving westward from Omaha in 1865, had effectively bisected the Great Plains herds into two distinct components by 1870.

In the north, substantial herds of bison still existed in 1880, concentrated from central Dakota Territory west to the Canadian prairies, the Rocky Mountains, and the Powder River Basin of Wyoming. The Northern Pacific Railroad had not yet penetrated the region, and hostile tribes of Dakotas, Assiniboine, Gros Ventre, Blackfeet, and Crow represented a significant threat to would-be buffalo hunters. However, at about the same time that military operations increasingly crushed Native resistance and progressively confined Native Americans to reservations, the Northern Pacific Railroad was pushing west into eastern Montana, the heart of the remaining bison range. Fort Benton in northeastern Montana became a hub of bison-hunting activities, with as many as one hundred thousand hides being shipped annually out of this location. Between 1870 and 1875 an estimated 2.5 million bison were being slaughtered per year, perhaps the

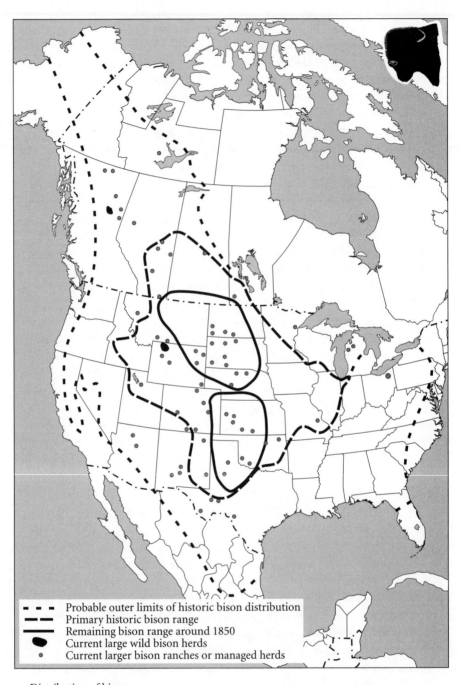

Probable outer limits of historic bison distribution
Primary historic bison range
Remaining bison range around 1850
Current large wild bison herds
Current larger bison ranches or managed herds

Distribution of bison

greatest mass killing of large mammals in history. By 1880 large numbers of hide hunters were converging on eastern Montana, after first dealing with the remaining animals in western Dakota Territory. In 1880 the last major kill by Dakota Natives in what is now the Cheyenne River Reservation resulted in a take of about two thousand hides. Three years later a group of white and Dakota hunters converged on the Black Hills to eliminate the last ten thousand animals that had taken refuge there. This was the last significant bison hunt by any of the Plains tribes (Haines 1995).

By 1884 the northern herds had also been decimated, and commercial buffalo hunts were rapidly abandoned after fruitless searches for large herds had become the norm. The military continued to try to eliminate the few remaining bison as effectively as possible, knowing that in doing so they were dooming the recalcitrant Native Americans still trying to live independently on the High Plains. Free ammunition was given to hunters at some army posts to help hasten the bison's elimination. Representative James Throckmorton of Texas summarized the prevailing national attitude by saying in 1876, "I believe it would be a great step forward in the civilization of the Indians and the preservation of peace on the border if there was not a single buffalo in existence." Wealthy hunters from the eastern United States and from Europe thus had to work ever harder to obtain a trophy, perhaps in hopes of even becoming famous for killing the very last wild bison. In 1886 W. T. Hornaday, the chief taxidermist of the U.S. National Museum, set out to obtain specimens for a large bison diorama. He finally managed to kill a total of twenty-six animals after months of hard hunting near Miles City, Montana. Six of these bison became the basis for the final museum group. Later, recognizing that the bison was virtually extinct, Hornaday organized the American Bison Society and convinced President Theodore Roosevelt to establish a National Bison Range in western Montana, in hopes of preserving a small breeding stock (McHugh 1972).

By the early 1900s, the focus of the national debate about the bison had changed gradually from how best to eliminate the troublesome herds to how some remnants could possibly be preserved for future generations to see and enjoy in national parks and sanctuaries. Many of the first small captive herds were started by capturing very young bison calves, which proved quite tame, especially if placed under the maternal care of cows or goats. Breathing in the nostrils of newborns also apparently generates an imprinting reaction, causing the animals to adopt the person who did the breathing as a parental figure. In this same way mother ungulates and their newborns learn to recognize each other individually in wild herds, marking an important aspect of the species' survival strategies for herding animals. Meriwether Lewis mentioned that on April 22, 1805, in what is now western North Dakota, a newborn bison calf attached itself to him and followed him until he got back on the expedition's pirogue. It has been estimated that about eighty calves probably obtained by such means produced the stock from which most bison surviving today have descended (Haines 1995).

In 1883 Frederick "Pete" Dupree, who had come to Fort Pierre in 1838, and his sons captured nine bison calves to begin a captive herd. By the time of his death in 1889 the herd numbered eighty-three head. In 1901 James "Scotty" Philip, a nearby rancher, bought the herd, and by 1918 it numbered five hundred animals, some of which were used to establish the Custer State Park herd.

Charles Jesse "Buffalo" Jones was a Kansas buffalo hunter who captured 14 calves in 1886, as the last herds were being eliminated on the Kansas plains. In only a few years he had developed a herd of 150 animals. Jones went bankrupt in the 1890s, and in 1893 his animals were all sold at auction and went on to stock new ranches in several states as well as provide a source of animals for recently created Yellowstone Park. Jones was eventually appointed the first game warden of the Yellowstone Game Preserve.

A very early Montana herd was begun in the 1880s by Michel Pablo, a man of mixed Hispanic and Piegan ancestry who lived on Montana's Flathead Reservation. He had acquired thirteen animals from Samuel Walking Coyote, a Pend d'Oreille Native American. Walking Coyote in turn had apparently acquired his bison as dependent calves following a Montana bison hunt. Pablo later obtained twenty-six of the bison from the herd that had been owned by Buffalo Jones, adding genetic diversity to his original Montana stock.

After Pablo's combined herd reached three hundred animals, the group was split. Half of the animals were retained by Pablo, with the other half going to the widow and children of Charles Allard, Pablo's former partner, who was also partly Native American. Some of these animals were soon sold to a Texas ranch, and the remainder were eventually sold by Pablo or the Allard family to various Canadian and U.S. parties. When the Flathead Reservation was opened to public settlement in the early 1900s, Pablo was forced to sell his entire herd, which then numbered about six hundred animals; all were acquired by the Canadian government.

Colonel Charles Goodnight began developing a personal captive bison herd in 1866, when he captured his first animal. He then gradually supplemented his stock with others. In 1902 he provided a few bulls to help restock the herd at Yellowstone Park, and in 1909 he also helped start the bison population at the National Bison Range in Montana. By the time of his death his herd numbered about fifty animals.

These relatively few herds of animals provided the nucleus for most of the nearly three hundred thousand bison alive today. Thus a very small handful of largely unrecognized individuals, including several Native Americans, were almost entirely responsible for saving the bison from total extinction.

Canada placed its few surviving wild bison under federal protection in 1889, and its only surviving unconfined herd of woods-adapted bison (the "wood buffalo") in the Northwest Territories soon numbered in the several hundreds (Haines 1995). The Canadian government also bought plains bison to introduce into Banff National Park after the park was established in 1899. In 1906 Canada bought nearly all of Pablo's remaining animals, which eventually resulted in 672 head being transported to western Canada. They were first shipped to Wainright, Alberta, and later placed in a newly established Buffalo National Park at Elk Island. After increasing in numbers and threatening to outstrip their range, over 2,000 of these bison were killed in the 1920s. Between 1925 and 1928 more than 6,600 were moved to Wood Buffalo National Park near Great Slave Lake, Northwest Territories. However, they were allowed to mix there with the local woods-adapted population, producing a group of animals having intermediate structural and ecological characteristics, and leaving only a small nucleus of animals wholly unaffected by plains genetic stock (McHugh 1972).

The loss of so many of America's bison to Canada finally stimulated the U.S. Congress to purchase some animals from Michel Pablo, initially to supplement the tiny surviving Yellowstone Park herd. This herd had finally been given federal protection in 1894, by which time its population had been reduced by poachers to about twenty animals. In 1908 Congress also established an 800-acre bison range in western Oklahoma, which was later enlarged to become the 59,000-acre Wichita Mountains Wildlife Refuge. In 1909 Congress also established the 20,000-acre National Bison Range in western Montana and released thirty-four bison that had been derived from the Pablo-Allard herd. After the army had abandoned Fort Niobrara Military Reservation in 1906, the land was set aside by President Roosevelt as a bird and wildlife preserve. In 1913 a small private herd of six bison was obtained locally and moved into the refuge. In the same year fourteen bison were placed in Wind Cave National Park in South Dakota's Black Hills, and thirty-six more were introduced into nearby Custer State Park. In 1917 six bison were placed in Sully's Hill National Game Preserve in northeastern North Dakota.

As of 2000, thanks to extensive state and federal efforts at restoration and a thriving private bison-ranching industry, an estimated minimum of 350,000 bison were living in North America. A nationwide survey by Sam Albrecht in 1999 revealed about 10,000 animals in U.S. publicly owned herds and more than 140,000 in private herds, plus 8,000 in Native American herds (Albrecht 2000). The Canadian population was then thought to have almost 100,000 bison in private herds and more than 3,200 in publicly owned herds in the late 1990s, counting both the plains bison and the so-called wood bison of northern Canada. Because of intermixing with animals of the plains population, only a small percentage of the northern herds are relatively "pure" wood bison, which – although considered a racially distinct population – is one that that Valerius Geist (1996) regards merely as a "phantom subspecies." Genetic studies by Gregory Wilson and Curtis Strobeck (1998) have measured the genetic diversities within and among the various present-day bison stocks and have confirmed that although no pure stock of wood bison now exists, these populations are sufficiently distinctive genetically to warrant separate management.

This free-living woodland population of Canadian bison was historically confined to Wood Buffalo National Park, Alberta, and Northwest Territories. It was classified as nationally threatened in 1988, after having been listed earlier as endangered. As of 1997 its population was judged to be about three thousand animals. The Wood Buffalo National Park herd numbered about two thousand head in 1994, down substantially from sixteen thousand in 1970 as a result of disease, slaughter, predation, and accidents, such as drowning after a reservoir was established nearby. The Elk Island National Park herd in Alberta numbered about nine hundred animals in 1993 and was partly composed of stock derived from the original Pablo–Allard plains stock in 1907, plus a separate group of wood buffalo stock obtained in 1965. A herd of a few hundred animals – a population derived from wood buffalo stock – also exists north of Great Slave Lake in the Mackenzie Bison Sanctuary. The Mackenzie and Wood Buffalo National Park herds are both relatively inbred but are now subject to the powerful effects of natural selection resulting from natural predation by wolves, which are preferably removing the less fit genetic stock.

Two other free-living northern British Columbia herds were established from the same genetic source of wood buffalo animals. One is a Nahanni–Liard River herd that still had fewer than a hundred animals in the mid-1990s; the other is a herd that escaped from a private bison ranch in the Pink Mountains and is also still fairly small. There are also a few bison at Hook Lake, outside Wood Buffalo National Park (Geist 1996). Because of genetic mixing between plains and wood buffalo stock as well as the additional intermixing of genes from the original "mountain" population of wild bison from Yellowstone, it is doubtful that any kind of taxonomic splitting of Canadian bison populations is now feasible.

The U.S. bison population is also substantial. In the United States, Yellowstone National Park has a herd that between 1968 and 1994 had grown from 400 to around 4,000 animals. About 450 were killed in 1996 when they strayed outside park boundaries, and a severe winter in 1996–97 caused a population crash. The 1997 summer population was about 2,000 animals. Nearby Grand Teton

National Park and Jackson Hole had another 300 or so head as of the mid-1990s. The Black Hills of South Dakota supports a managed herd of some 1,500 head in Custer State Park plus about 350 more in adjacent Wind Cave National Park and additional animals at Badlands National Park. There are managed herds of about 500 head at Fort Niobrara National Wildlife Refuge in Nebraska and around 450 at Theodore Roosevelt National Park in North Dakota. Another 600 bison reside in the Wichita Mountains Wildlife Refuge in western Oklahoma, and 300 to 500 bison are maintained on the National Bison Range in Montana. A similar number are at Badlands National Park, South Dakota. There is also a free-ranging herd of plains bison (about 300 head as of 1998) along the northern edge of the Alaska Range in the Kuskokwim River drainage (the Farewell Station herd), which is managed for sport hunting by the Alaska Department of Fish and Game. Of the many private bison ranches, Ted Turner's operations in several states alone support nearly 10,000 animals, and several other private operations have more than 1,000 animals each. Several Indian reservations on the High Plains, including the Cheyenne River, Pine Ridge, and Lower Brule reservations in South Dakota, are building up their herds to substantial sizes as well. As of 2000, forty-six tribes were holding 8,000 to 10,000 bison, organized through their Intertribal Bison Cooperative system.

Compared with a 1972 estimate of 30,114 bison in all of North America, bison numbers have increased roughly tenfold in only thirty years, an incremental rate of increase of about 15 percent per year. Bison ranching will be discussed further in chapter 10.

My friend,
They will return again.
All over the earth,
They are returning again.
Ancient teachings of the Earth,
Ancient songs of the Earth,
They are returning again.
– *Tshunca-Uitco (Crazy Horse)*

THE LIFE AND BEHAVIOR OF BISON

It is a common mistake, and a great underestimation, to think of bison as little more than big, woolly cattle. Unlike cattle, whose ancestors evolved in a subtropical environment where water was readily available and freezing temperatures unheard of, the modern bison evolved about ten thousand years ago near the end of the Ice Age in the high-latitude steppes of northern North America. Its immediate ancestors were even larger and stronger arctic-adapted mammals with massive horns and high shoulder humps, whose associated long and powerful muscles helped extend their stride when running. Modern-day bison have somewhat less well developed shoulder humps but nevertheless have been known to gallop 35 miles per hour over a distance of half a mile. Early bison probably needed little water and were adapted to feeding on low-stature arctic plants with limited productivity. As a result modern-day bison are inherently prone to wandering, being almost constantly on the move, and thus allowing the short plants of the high plains to be grazed only lightly before the animals proceed to new pasturelands.

Like other herd animals, bison's socialization abilities are important. Mature females are the accepted leaders of herds; when males become sexually mature at three or four years of age, they leave the protection of the large herds of females and immature males. These bulls then associate separately with other adult males, returning to mix with the females only during the rutting period in early fall. Males are prone to gather in small groups where food is abundant, whereas female-led groups favor level, hard grounds where visibility is unobstructed and running to escape from wolves or other dangers is more feasible.

Compared with females, bison bulls are substantially larger and stronger, and they have more massive heads and horns as well as much thicker and tougher skins. They also have great matted cushions of hairs up to two feet in length in front of their thick skulls, which may help to cushion blows during head-to-head fights with other males. Their powerful neck muscles and upward-pointing horns allow them to push and smash their way through crusted snow to get at grasses hidden far underneath. The thick hair around the shoulders that is so evident in rutting males probably serves to visually signify male virility, in the same way a male African lion's mane does. The length and condition of this hair accurately reflect the animal's age and health and, at least indirectly, may also reflect the quality of its foraging habitat. After the autumn rut much of this "display hair" is shed rather than being retained through winter as might be expected if its primary function were to provide cold-weather insulation (Geist 1996).

It is known that bison can survive for much longer periods without water than cattle can, although they will often walk several miles to a water hole if one is available. Such rivers as the Platte and Republican were favorite fall and spring staging areas for migrating bison. One Native American name for the Republican River was "Turd River," because of its extensive use by bison. By May and June the Platte was sometimes almost dry, certainly in part because of its use by bison. Bison may drink 5 to 10 gallons of water per day and consume or trample about 30 pounds of forage. Bison's water needs are probably fairly similar to those of elk, which at least originally also were primarily High Plains mammals. Cattle require three to five times as much water as elk but weigh about twice as much and generate about 20 pounds of urine and up to 50 pounds of manure per day (Jacobs 1991) as well as about 400 quarts of methane daily, a major greenhouse gas. Bison conserve a considerable amount of their urea in their rumen, which produces a high concentration of bacteria and protozoa and allows for better fermentation and digestion of high-cellulose vegetation.

Bison also have unusually broad incisors and very large molars and premolars, adapting them to clip and chew short grasses that are high in fiber and silica (Geist 1996). The High Plains grasses and forbs also tend to be rich in diverse mineral salts, which may be valuable for building massive bone structure and resistance to fracturing. Also unlike cattle, bison favor foraging on upland grasses over wide areas. Cattle concentrate on taller grasses and forbs growing near water, producing massive local overgrazing, trampling of vegetation, and soil compression. In spite of the ecological advantages of bison on grasslands, there are roughly 98 million cattle in the United States compared to about 150,000 bison, roughly 650 times as many animals. The standing biomass of cattle in the United States relative to that of humans (assuming an average weight of 750 pounds for cattle and 100 pounds for humans) is about 3.6 times greater. An alien visitor to America might justifiably conclude that we represent the United States of Cattle. Worldwide, there are about 1.3 billion cattle, as compared to 6 billion humans, again representing a greater cattle than human world biomass as well as adding an unbelievable load of fluid and solid wastes, methane, water pollution, and stress to the planet's already weakened ecosystems.

Bison calves are born in early summer; typically a single calf is born per pregnancy, and the calves are relatively large (30 to 70 pounds) at birth. They are fed on milk that is rich in fat, sugar, and proteins, and they grow rapidly. Their cinnamon red coat color is conspicuously different from adult coats, and they carry this distinctive color for about six months. Like other ungulates they are precocial and able to run well soon after birth, a valuable asset when prairie wolves roamed the plains. Like other ungulates bison are most vulnerable when frightened into fleeing full tilt from predators, for it is then that weak or slow animals are most likely to be recognized, separated from the herd, and set upon individually. Bison are unlike muskoxen in that they do not form protective circles around females and young; however, small groups of males may associate defensively in loose cooperative stands.

The annual rut begins in late summer and lasts through early fall, roughly between late July and early September. The northern populations start rutting later than the southern ones, with the season presumably timed so that the

Fig. 1. Bison, adult males, head-to-head fighting during fall rut.

Fig. 2. Bison, adult male bellowing. The tail position sketches above indicate (left to right) mild excitement, defecation, great excitement, and combative postures (after Steelquist 1998).

young will be born the following year in early summer, their optimal time for birth. At this time each male tests his strength against other competing bulls, with the encounters ranging from fairly tame episodes of head-to-head pushing contests (Fig. 1) to all-out frontal attacks in which the bulls crash into each other, heads held low and sharp horns pointed directly forward toward the opponent. Severe wounds may result from these contests, which determine each male's prospects for gaining access to sexually available females. Dust kicking, horning, and wallowing by males are also common behaviors during the rut, as is bellowing. Dust kicking is a visual sign of aggression, as is lifting the tail. Horning the ground and jabbing, thrashing, or butting nearby trees are also common, seemingly as a form of redirected aggression toward nearby objects as a way to handle and release frustrations. Wallowing and rubbing of trees are also common and seem to have scent-marking functions that might be related to the priming of estrus in females, according to R. Terry Bowyer (1998) and others. Wallowing probably provides some protection to the animal by giving it a coating of mud, but wallowing behavior is most common among males, especially in front of females and rival males.

Bellowing (Fig. 2) is common among rutting males and varies from a soft purring to a great roar that is a direct challenge to all rivals. It is uttered with the male's mouth open, his upper lips retracted, his tongue extended, and his underparts visibly compressing with each breath as air is expelled. The sound may carry up to 3 miles under favorable conditions, often eliciting a response from some distant, unseen bull (McHugh 1972).

Females that are in heat are repeatedly "tested" by males, who closely follow behind such females, forming "tending bonds" (Fig. 3). During this following behavior, which may last for days, the male often protrudes his tongue toward the female while extending his neck forward and holding his head level (Fig. 4). He repeatedly tests her urine by licking it or closely inspecting the cow's now swollen genitals while retracting his upper lips in a "lip-curl" display. He then retracts his tongue to bring it into contact with a palatal structure that is a sensitive chemoreceptor, permitting the male to detect the relative state of estrus in the female. A female ready to be mated will allow herself to be licked and nudged prior to copulation. Soon the male will lose interest in the just-mated female and go off in search of other estrus females (Geist 1996).

Before giving birth, the pregnant females move away from the large herd and form nursery groups consisting of adults, young calves, and some immature animals still socially attached to their mothers. After birth the calf divides its time between nursing with the mother and playing with other calves (Fig. 5). Gradually, during autumn, the nursery groups disband and the herds reassemble to prepare themselves for the hardships of the long northern winter.

Fig. 3. Bison, adult male posture when "tending" estrus female.

Fig. 4. Bison, adult male extending tongue toward estrus female.

Fig. 5. Bison, female nursing calf.

By about age seven years, males are past their prime and become increasingly unable to compete effectively during the rut. Such males may become loners, living out their remaining time in relative seclusion or associating quietly with a single companion. It is thought that the average lifespan of bison is no more than fifteen or sixteen years, but some individuals living in protected environments may survive beyond thirty years. A few animals have been documented to live at least thirty-eight years.

What is life? It is the flash of a firefly in the night. It is the breath of a buffalo in the wintertime. It is the little shadow which runs across the grass and loses itself in the sunset. – *Crowfoot, Blackfoot warrior, 1890*

3

Prairie Dogs and the American West

Little Towns on the Prairie

In this plain and from one to nine miles from the river or any water, we saw the largest collection of the burrowing or barking squirrels that we had ever yet seen; we passed through a skirt of the territory of this community for about seven miles. – *Meriwether Lewis, June 5, 1805, near the Marais River, Montana*

The immense numbers of animals in some of these towns, or warrens, may be conjectured from the large space which they sometimes cover. . . . Estimating the holes to be at the usual distances of about twenty yards apart, and each burrow occupied by a family of four or five dogs, I fancy the aggregate population would be greater than any other city in the universe. – *Capt. Randolph Marcy, June 26, 1852, near the Red River, Texas*

If the numbers of bison seen by early explorers of the West seemed uncountable, then the numbers of prairie dogs they sometimes reported were simply astronomical. The quote by Captain Marcy just noted came about as he was leading an exploratory military survey of the Red River (also known in Spanish as the Rio Colorado) and referred to a prairie dog town that he estimated to be some 25 miles in continuous length. Assuming it to be just as wide, he estimated its total area as 625 square miles, or 896,000 acres. Also assuming a modern-day population density estimate of about ten animals per acre, nearly 10 million animals may have lived in that single town, more than the total number of humans living in any city then on earth. Whoever first said that the meek shall inherit the earth perhaps had prairie dogs (before their persecution) in mind.

Other early observers were also greatly impressed by prairie dog numbers. C. H. Merriam, the first director of the U.S. Biological Survey (later to become the U.S. Fish and Wildlife Service) believed that prairie dogs in 1900 occupied some 63 million acres of the western states (Merriam 1902). Vernon Bailey, a biologist for the U.S. Biological Survey and also Merriam's brother-in-law, reported a prairie dog colony north of San Angelo, Texas, some 100 miles wide and about 250 miles long (Bailey 1905). He thought there might be 400 million animals on this single area of about 16 million acres, a probable overestimate.

Bailey (1931) also thought that there might be 6.4 million prairie dogs in New Mexico's Grant County alone (an area totaling 2.5 million acres, representing an average of about 2.5 animals per acre) as of 1931.

Recent speculations by various writers have suggested that as many as 5 billion black-tailed prairie dogs may have inhabited the shortgrass and mixed-grass High Plains region during presettlement times. This high a figure seems unlikely, however, and would require a population density of about 15 animals per acre over the entire half-million square miles of original shortgrass and mixed-grass prairies. A more conservative estimate of usual colony densities might be about half that figure, or 7.5 animals per acre, but the idea of even 2.5 billion prairie dogs on the High Plains is still mind-boggling and assumes that the entire High Plains region was once fully occupied. A recent review paper by Craig Knowles, Jonathan Procter, and Steven Forest (2002) suggested that, prior to European-American settlement, black-tailed prairie dogs probably occupied 2 to 15 percent of large landscapes within the Great Plains. Taking 8 percent of the western Great Plains as a rough compromise estimate of occupied territory, we are still left with a possible presettlement population of some 200 million black-tailed prairie dogs, a population reached by humans in the United States only during the past few decades. With an average adult weight of 2 pounds, that translates into about 400 million pounds of potential food for predators. By comparison, the 30 million bison presumably once occurring across the grasslands and woodlands of North America would have represented at least 30 billion pounds of living biomass.

The black-tailed prairie dog (Fig. 6), the best known of North America's five species of prairie dogs, has played a role in Native American mythology and more recent Western folklore almost as great as that of the Native American bison. To the early French explorers the prairie dog was known as the *petit chien*, or little dog, a name that Captain Meriwether Lewis was aware of when he first encountered and carefully described the animal. Lewis instead chose to call it the "barking squirrel," but later he also sometimes referred to it as the "burrowing squirrel." However, other party members, including Captain Clark, John Gass, and Patrick Ordway, commented on the "prairie dog" in their own written versions of that species' discovery. A few decades later the species came to be known as the "prairie dog" or "prairie marmot squirrel" by such explorer-biologists as John James Audubon and Prince Maximilian of Weid. Audubon mentioned that trappers and Native Americans referred to the prairie dog's colonies of up to hundreds of families as dog towns or villages, which are "intersected by streets (pathways) for their accommodation, and a degree of neatness and cleanliness is preserved." Audubon also noted more accurately

Fig. 6. Adult black-tailed prairie dog.

that burrowing owls live among the towns, as do rattlesnakes and other reptiles (Audubon and Bachman 1854).

As it is the only U.S. species of prairie dog to have a black-tipped tail, the additional descriptor "black-tailed" was eventually also added to the species' English name. Its formal current Latin name, *Cynomys ludovicianus*, translates as the "dog-mouse of Louisiana Territory." The animal's formal scientific description (its genus originally having been designated as *Arctomys*) was given in 1815, probably on the basis of specimens brought back by the Lewis and Clark expedition of this newly acquired territory. Early in the expedition Lewis had sent back from Fort Mandan several living animals, as well as two skinned specimens, to President Jefferson. One of the living prairie dogs actually survived the long trip and for a time was exhibited alive in Charles Peale's Philadelphia Museum, then located in Independence Hall.

PRAIRIE DOG SPECIES AND DISTRIBUTIONS

Along with the bison and coyote, prairie dogs are as much a symbol of our mythic West as are the gold-seeking hermit and the lonesome cowboy. Perhaps their onetime sheer abundance is what attracted early human attention, but their talkative nature and their puppylike barks also appealed to early explorers and describers of the western scene. In part, too, it was perhaps the erect, somewhat humanoid posture of alert prairie dogs, and their tendency to interact as a small community, that made humans identify with this rodent more strongly than is typical of human reactions to ground squirrels generally. Visitors to national parks and other wildlife preserves are just about as likely to stop along roadsides to watch prairie dogs as they are to observe deer, elk, or other much larger wildlife.

The black-tailed prairie dog is the most widely distributed of all the prairie dog species (Map 5), although its presettlement range and populations have been severely reduced as a result of a century of unremitting efforts by ranchers and farmers to eliminate the species (Table 1). The distributions and biologies of the other four species are of equal interest. These species include the Utah prairie dog, the Gunnison's prairie dog, the white-tailed prairie dog, and the extra-limital Mexican prairie dog, which is now well isolated in the southern Chihuahuan Desert in northeastern Mexico.

These four species now all have barely overlapping but boundary-sharing distributions. The white-tailed and Gunnison's prairie dogs meet or nearly meet the black-tailed species at the western edge of the black-tailed range, a geographic division that is generally correlated with the Continental Divide. The most extensive area of actual contact between the black-tailed and the white-tailed species is in Wyoming, where these two are in apparent contact in several central or western counties, though they actually tend to remain separated by differing elevation preferences. The white-tailed also occurs to the north of the Gunnison's species; these two populations effectively are separated by the Gunnison River in west-central Colorado and eastern Utah. These two species have limited geographic contact in western Colorado, but no clear evidence of past or present hybridization exists.

The Utah prairie dog (Fig. 7) is more completely isolated from both the white-tailed and the Gunnison's prairie dog in the mountain valleys of southwestern Utah. It presumably evolved as an offshoot of the Gunnison's or the white-tailed species at some earlier time when climatic conditions in the western Rocky Mountains were less desertlike. At that time the ancestral species' range was probably much broader and presumably was in contact with one of these more easterly types. It is likely that the Utah prairie dog's nearest relative is the white-tailed species, whose range fairly closely approaches its own in northeastern Utah. The Utah prairie dog is now extremely rare, having been listed as endangered in 1973. Its status was upgraded in 1984 to threatened, as a result of local political pressures for allowing control measures. It is currently (as of 2003) again being considered for endangered status.

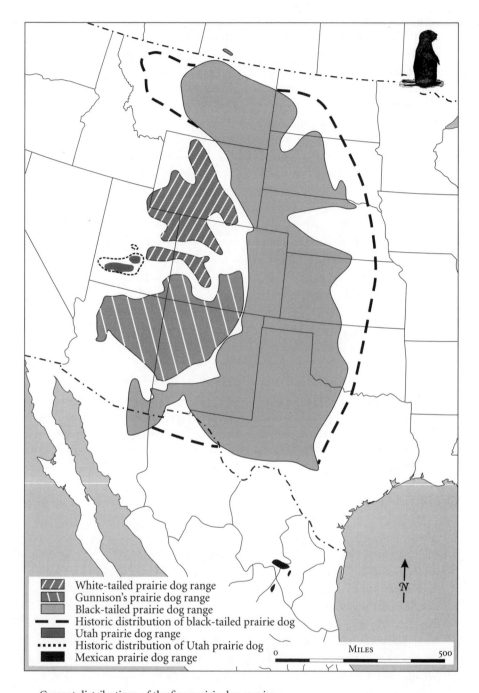

Current distributions of the five prairie dog species

Table 1. Estimated historic and recent U.S. black-tailed prairie dog populations

YEAR		ESTIMATED TOTAL COLONY ACREAGE
Historic		368,308,727 available habitat [a]
1870		116,000,000 available habitat [b]
2002	Colorado	631,102 (44,000 [b])
	Texas	197,000 (22,500 [b])
	South Dakota	160,000 [c] (142,272 [d])
	Kansas	130,321 (36,000 [b])
	Wyoming	125,000 [c] (329,078 [d])
	Montana	90,000 (65,000 [b])
	Nebraska	80,000 (81,441 [d])
	Oklahoma	64,214 (8,500 [b])
	New Mexico	60,000 [c] (60,000 [d])
	North Dakota	20,500 (34,555 [d])
	Arizona	0 [b, e]
Total recent U.S. estimates:	High end [f]	1,558,337 [a] available habitat
	Low end [g]	823,346

[a] Luce 2003.

[b] Estimates of National Wildlife Federation (1998).

[c] Incomplete survey.

[d] Late 1990s estimates of active colonies by Sidle et al. (2001). (A 2004 estimate for South Dakota was 359,000 acres, and a 2003 estimate for Nebraska was 137,000 acres.)

[e] Reintroduction efforts are planned.

[f] The high-end estimates for Colorado have been seriously questioned (Rosmarino, Robertson, & Crawford 2003), and it is likely that most states have also recently inflated their acreage estimates for political reasons. No total estimates of individuals are available, but at a high-end average of 10 animals/acre, the total U.S. population might have approximated 3.6 billion historically, 1.1 billion animals in 1870, and 8.2–16 million in 2000, in addition to Mexican (about 1.5 million) and Canadian (under 20,000) populations (see also Table 6).

[g] Late 1990s surveys of occupied habitats and 1998 National Wildlife Federation estimates.

The Utah, white-tailed, and Gunnison's prairie dogs are believed to represent a group of three closely related species, while the black-tailed and Mexican prairie dogs represent another, slightly divergent lineage. Chromosomal and biochemical data suggest these relationships, indicating that the black-tailed and Mexican species are closely related and that the Utah and white-tailed prairie dogs form a separate but closely related group. John J. Pizzimenti (1975) has further suggested that gradual mountain uplifts separated the ancestral plains populations from those in the mountains and western slopes. He regarded Gunnison's as the most primitive species, with the Mexican prairie dog as a

Fig. 7. Adult Utah prairie dog. After various sources.

relict population of the black-tailed, and the Utah prairie dog as a similar relict isolate of the white-tailed. He suggested that the prairie dogs evolved from ground squirrels in the southern Rocky Mountains during Pleistocene times, gradually extending both east and west from this point of origin. However, recent DNA studies suggest that the prairie dogs as a group probably evolved from desert-adapted ground squirrels, an idea that seems more in line with the general origin pattern of the High Plains fauna and flora.

In the black-tailed and the Mexican prairie dogs the tail is tipped with black. In the Mexican species this dark color covers nearly the entire outer half of the tail, but in the black-tailed it is limited to the outer third. In the Gunnison's prairie dog the outer half of the relatively short tail has a gray center, which is bordered and tipped with grayish white. In the other two species the outer half of the tail is white, but the body color of the white-tailed species tends to be generally buffy to grayish overall, rather than rich cinnamon as in the Utah. The Utah and the white-tailed prairie dogs both tend to have a conspicuous blackish spot above each eye. The Mexican species is noticeably smaller than the black-tailed, and the Gunnison's is the smallest of all. A large Gunnison's may weigh about 2 pounds, whereas a large black-tailed may weigh as much as 4 pounds.

Fig. 8. Adult Gunnison's prairie dog.
After various sources.

The presumably primitive Gunnison's prairie dog (Fig. 8) has a considerably smaller number of chromosomes than the others as well as a behavior pattern more like that of the ground squirrels than the more typical prairie dogs. Gunnison's populations tend to be small and patchy, and their mounds and burrow systems are also generally more similar to those of typical ground squirrels than to the highly social prairie dog species. The small mound at the edge of the burrow is not well developed or especially modified to serve as an elevated lookout station. The excavated burrows tend to be shallower than in black-tailed prairie dogs, often only about 3 feet deep. They may have several entrances, as well as several lateral tunnels that do not reach the surface. Territoriality is not well developed among the males, and social relationships seem to center on mother-offspring connections. Adult males may live somewhat apart from breeding females during summer, and yearling males may live closer to the females. Population densities average only about one or two adults per acre (Tables 2 and 3).

The Gunnison's prairie dog has declined by over 90 percent throughout its range, as a result of poisoning, shooting, sylvatic plague, and habitat destruction. Plague has recently devastated prairie dog populations across large areas of northern Arizona, and habitat destruction has resulted in acreage reduc-

Table 2. Recent acreage and population estimates for three prairie dog species

STATE	WHITE-TAILED	GUNNISON'S	UTAH
Arizona	–	106,000	–
Colorado	125,766	85,795	–
Montana	120	–	–
New Mexico	–	No information	–
Utah	97,786	3,678	Threatened
Wyoming	340,000	–	–
Total acres	600,000–800,000	200,000–335,600	Not reported
Total estimated animals	1.2–1.6 million	1.0–1.7 million	4,217*

Source: Table data is based on Knowles (2002), except for the Utah prairie dog. Other survey estimates are for around the year 2000; these surveys probably were incomplete, accounting for differences between listed individual state acreages and estimated total overall acreages. The estimated occupied range of the Mexican prairie dog in 1992–93 was given by Treviño-Villarreal and Grant (1998) as 478 square kilometers (about 110,066 acres).

*Unpublished 2002 total population estimate from the Utah Division of Wildlife Resources. Total area of species was estimated at 75,000 acres in 1984; plague epidemics since then have caused serious declines in populations and acreages. Listed as nationally endangered in 1973; changed to threatened in 1984. Government translocation programs have resulted in a serious source of mortality in recent years.

tions in Arizona and New Mexico. Uncontrolled shooting of Gunnison's prairie dogs occurs in Colorado, and increased oil and gas development is eroding the species' little remaining habitat. In February 2004 a coalition of scientists, private landowners, religious organizations, conservation and animal protection groups, and other concerned citizens filed a petition to list the Gunnison's prairie dog under the federal Endangered Species Act. As of late 2004 the petition had yet to be acted upon.

The white-tailed prairie dog has a degree of sociality similar to that of the Gunnison's, averaging about 1.3 adults per acre, while burrow openings average about 22 per acre. According to Tim Clark (1977), the burrows have associated mounds composed of excavated subsoil. However, the mounds are not shaped, packed, or worked in with topsoil, as they are with black-tailed prairie dogs. One burrow excavated by Clark was about 11 feet long and reached a maximum depth of more than 3 feet. Another longer and more complicated burrow was not fully excavated but was plugged in two places with prairie dog feces. Neither was similar to the complex burrow system of the black-tailed prairie dog. Likewise, the social system of the white-tailed species is much like that of the Gunnison's; the primary social unit is a short-term association between lactating females and their dependent young. Playing among the young and "kissing" (nose-to-nose contact) between females and their young or between adults during the breeding

Table 3. Summer prairie dog and burrow densities

SPECIES	ANIMALS/ACRE	BURROWS/ACRE	REFERENCE
Black-tailed	22.7	53.1	Davis 1966
	8.5–12.1	68.5	O'Melia 1980
	6.3	41.9	Tileston and Lechleitner 1966
	2.0–7.4	27.0	Koford 1958
	5.4–15.0	54.8	King 1955
	1.6–12.4	41.5	Knowles 1982
White-tailed	0.3–2.0	10.5	Clark 1977, Clark et al. 1982
	2.5–3.2	11.3	Tileston and Lechleitner 1966
	2.3–6.5	15.6–43.9	Biggins et al. 1993
Gunnison's	4.9	23.1	Fitzgerald and Lechleitner 1974

Note: See also Knowles (2002) for additional estimates.

season are among the few well-documented social behaviors, which also occur in the other prairie dogs species. Some obviously sexual and aggressive interactions also occur, as well as an integration of colony behavior by vocal and visual signaling. These signals include a fully standing alert posture and an upright-alert posture while sitting (Fig. 9). When this posture is assumed in conjunction with calling by one animal, similar alert postures are assumed by other members of the colony. The upright-alert posture is also used by males during territorial calling, but in this situation the head is held at a distinctly elevated angle. In July 2002 the Center for Native Ecosystems, Biodiversity Conservation Alliance, Southern Utah Conservation Alliance, American Lands Alliance, and Forest Guardians petitioned the federal government to list the white-tailed prairie dog as threatened or endangered. The U.S. Fish and Wildlife Service had not responded to this petition as of the summer of 2004, owing to various legal delays. The petition seeks Endangered Species Act protection for a species that has been eliminated from over 90 percent of its historical range but nevertheless is under intense political pressure to be removed from the list of candidates for federal protection.

The Utah prairie dog is now confined to a few mountain valleys in southern Utah, where its range has been reduced by more than 90 percent in the past half-century. Because of poisoning efforts by state, federal, and private interests, its population has now been reduced to a few thousand animals concentrated in three counties (Iron, Garfield, and Wayne). It is loosely colonial, with population densities ranging from only about one adult per acre to about thirty per acre, the latter in areas of lush but not too tall vegetation. Its burrows are like those of the white-tailed, in that they have a mound at the entrance that is composed

Fig. 9. White-tailed prairie dog, adults in upright-alert and upright-extended postures. Upper right sketch shows territorial calling posture of male. After illustrations in Clark 1977.

only of subsoil and is haphazardly formed. Single-entrance burrows with small mounds are most common, but old burrow systems may have five or more entrances, with associated mounds up to 2 feet high and 10 feet wide. Like the white-tailed, the burrow of the Utah prairie dog descends at an angle of about 45 degrees from the opening, but the subterranean details of the burrow system remain unstudied. These relatively high elevation animals hibernate through the winter, as do white-tailed prairie dogs, in contrast to the black-tailed and the Mexican prairie dogs, which do not hibernate, and to the Gunnison's, which hibernates only intermittently.

The Mexican prairie dog is now extremely rare and has been classified as endangered by the U.S. Department of Interior. It may have originated as a subpopulation of the black-tailed prairie dog, whose range extended much farther south during the Wisconsin glaciation. It now is limited to a small area centering near the intersection of the states of Coahuila, Nuevo Leon, Zacatecas, and San Luis Potosi, concentrating in valleys, prairies, and basins at about 5,000 to 7,000 feet above sea level. The animals favor areas of mature desert vegetation, well covered with grasses and herbs, and occur in colonies of up to hundreds of individuals. What little is yet known of the species' behavior indicates that it is as complex as the black-tailed prairie dog's. A well-developed mound is present at the burrow's entrance, which serves as a lookout post and marks a burrow that descends in a steep spiral for about 3 feet before leveling off. Several males and females and their offspring live in the same burrow, and inactive burrows may also be used by burrowing owls and spotted ground squirrels.

THE BLACK-TAILED PRAIRIE DOG

The black-tailed prairie dog is by far the best studied of all the prairie dog species, simply because of its wide distribution and very high populations as compared with all the others. As such it provides a general model for prairie dog ecology and behavior, although its social life is the most complex of all, and it is the only one of the U.S. species that does not normally hibernate, despite having a range that extends farther north than any of the others. This lack of hibernation may in part reflect its large body size, which allows it to withstand cold temperatures better than smaller species. However, prairie dogs may remain underground for several days during very cold weather; for example, John Hoogland (1994) noted that sixteen small individuals in poor condition that he monitored did not appear above ground from perhaps as early as November until March.

The black-tailed species is also unique in that tall plants growing within the ward are clipped. These plants are not eaten; such clipping likely provides an unobstructed view of the immediate vicinity, thus reducing the chances of undetected approach by ground predators. The black-tailed prairie dog's burrows often have multiple entrances, usually two but at times up to six. Some unused burrow entrances may be plugged with dirt or debris, and extra entrances to the nursery burrow may be plugged by females caring for unweaned young. Burrow entrances that have been used by predators may also be plugged, at least temporarily.

The burrow entrances of this species are typically somewhat domed or variously rimmed. Burrows that are rimmed with dirt all around the entrance certainly help to prevent flooding, and mounded entrances serve well as lookout points. Burrow entrance features may also help with ventilation. When a burrow has a relatively low entrance and a second entrance at a higher level, a partial vacuum is created that results in a directed airflow through the burrow as warm air moves upward to the higher entrance, thus improving burrow ventilation. Burrows with sharp rims at their ends have been found to function more effectively as exit points for airflows than do mounds with rounded tops (Vogel, Ellington, and Kilgore 1973). Black-tailed prairie dog burrows may be as long as 100 feet and as deep as 16 feet, a situation that – without proper ventilation – could otherwise result in serious dead-air problems for the sleeping animals.

Black-tailed prairie dogs, like all other prairie dogs and ground squirrels, are active only during daylight hours, spending more than 95 percent of their time above ground during daytime in good weather. Perhaps the northern extension of the black-tailed range is helped by the long days associated with these higher latitudes, providing more hours per day to put on a layer of body fat. However, copulation occurs underground, as does most nursing (Hoogland 1994).

Like the other prairie dog species, the black-tailed is almost entirely a vegetarian, although it occasionally also eats insects. Cannibalism by lactating females of other females' unweaned young also occurs, at least in the black-tailed. Black-tailed prairie dogs sometimes eat the old or even fresh scats of bison. Perhaps, as in rabbits that regularly consume their own droppings, some vitamins or other important nutrients are extracted from these excreted materials.

Black-tailed prairie dogs generally reach sexual maturity at two years, although some males have been found not to reach this milestone until their fourth year. About a third of the females studied by John Hoogland (1994) were sexually active as yearlings. Hoogland also recorded his oldest females as eight years, versus five years for males. His study registered about a 50 percent mortality rate between first emergence from the burrow and the end of the first year of life.

Foods and Foraging Behavior

Foods of the black-tailed prairie dog have been the focus of many studies. One done by J. A. Fagerstone (1979) in the Buffalo Gap National Grassland of South Dakota is representative. There, grasses formed the dominant part of the diet from spring to early winter, with prickly pear cactus dominating the diet in February, perhaps as an adaptation to water stress. Plants were eaten as soon as they began to grow, with cool-season grasses consumed during the spring and summer, and warm-season grasses, such as buffalo grass and blue grama grass, later in the summer. Typically, prairie dogs and cattle prefer much the same foods, and thus their activities may decrease the grasses considered beneficial for cattle. However, plant species diversity and plant protein content are often greater on prairie dog colonies than off them, which may help explain why they are so attractive as grazing sites for bison, pronghorn, and other larger grazers. One reason large ungulates favor grazing in prairie dog towns is that their burrowing activities move subsoil minerals around, tending to bring them closer to the surface and thus making them more available to shallow-rooted plants.

In another study, A. C. Lerwick (1974) found that prairie dog grazing exerted selective pressure against blue grama but favored the increase of buffalo grass as well as invasion by annuals. During a drought, cattle increased their use of forbs, while prairie dogs relied more heavily on grasses, and neither consumed buffalo grass if other foods were available. The favorite food for both species was blue grama. Research by Pfeiffer, Reinking, and Hamilton (1979) has shown that prairie dogs can survive for long periods (at least six weeks) without food or water during summer, although their weight may drop by 50 percent. In winter, starved prairie dogs exhibit a metabolism similar to that of black bears during winter sleep.

Competition with other grazers, especially cattle, has been the subject of many studies. Research by O'Melia, Knopf, and Lewis (1982) is representative. The researchers found that the presence of prairie dogs decreased forage availability and utilization by cattle but did not significantly reduce steer weight gains in either of two years. They suggested that improved soil fertility, nutrient cycling, and subsequent changes in forage quality at least partly compensated for reduced forage availability. Prairie dog pastures also supported an arthropod biomass (mostly grasshoppers) three times greater than control pastures, probably providing increased supplies of insect foods for insect eaters such as birds and small mammals.

Of the various mortality factors affecting prairie dogs, predation is generally a minor factor, perhaps as a reflection of the excellent antipredator communication system present, especially in black-tailed prairie dogs. In Tim Clark's (1973, 1977) studies of white-tailed prairie dogs, predators were known to account for only four deaths, three by badgers and one probably by a golden eagle. Other studies of this species have also implicated badgers as primary predators. However, a flea-transmitted disease in the form of sylvatic plague (also called rodent plague) killed nearly 85 percent of Clark's study population during four months. Clark believed that the white-tailed prairie dog's relatively looser social organization and higher migration rate may account for the fact that sylvatic plague seems to have greater impact on white-tailed prairie dogs than it does on black-tailed prairie dogs.

However, plague can readily eliminate entire colonies of black-tailed prairie dogs soon after it is initially introduced, perhaps because of the species' high level of sociality and generally high densities. Because of the high sensitivity of prairie dogs to plague (the animals often die within a few days of infection), prairie dogs do not pose a significant disease threat to humans. Rodents such as rock squirrels, which are more resistant to the disease and often live close to humans, are much more likely to be carriers that could infect other species. The form of this bacterial disease affecting humans directly from flea bites is bubonic plague; inflected humans can transmit the disease directly to others in the form of pneumonic plague, through airborne transmission. Both forms of the plague are fatal if left untreated.

In John Hoogland's (1994) study site, where plague did not occur, infanticide by the prairie dogs themselves was a major mortality factor. In sixteen years of study Hoogland documented only twenty-six predator events on his study area: ten performed by bobcats, six by prairie falcons, three by coyotes, three by golden eagles, two by probable Cooper's hawks, and one by a badger.

Social and Reproductive Behavior

The social organization of a black-tailed prairie dog town is surprisingly similar to that of a large human community. Based on terminology first used by John A. King (1955), the species' towns (colonies) are initially subdivided into smaller units called wards, which are largely determined by topographic or vegetational features. Each ward is in turn subdivided into coteries, which are comparable to the clans or harems of other prairie dog species, the upper limit of their social organization. Each coterie in turn consists of one or, rarely, more adult males, several adult or immature females that are closely related to one another, and additional juveniles of both sexes. Unlike other prairie dogs, black-tailed typically require two years to become sexually mature, improving the possibility of large family groupings developing without serious social strife.

Each coterie occupies about an acre, and its limits define both the home ranges and the territorial boundaries of its members. Coterie boundaries do not normally overlap, and their limits are strongly defended. Densities within wards can range from 3.5 to 12 adults and young per acre, a density level rarely attained by any other prairie dog species. Such high densities reduce the chances of any single animal being taken by predation but increase the possibility of disease and parasite transmission as well as competition for food.

One of the most socially organized of all rodents, the black-tailed prairie dog has a necessarily complex pattern of social communication signals. Its vocal signals include a doglike barking or repeated "chirk" call, which is thought to alert others to a predator's presence. Such calls will stimulate other prairie dogs to search for the predator and also respond with the same alarm call. An approaching predator will quickly send all prairie dogs into their burrows, so that an effective and quickly broadcasted alarm system is one of the obvious benefits of social living, perhaps serving as a counterbalance to the disadvantages caused by so many possible prey individuals living in a small area and attracting predators (even if the danger to an individual animal is reduced statistically by being part of a group). Besides the alarm call the black-tailed has a possibly defensive "chuckle" note, as well as growling, snarling, and chatterlike calls used in threat, and a purring call that is associated with mating.

At least eleven different vocalizations are known from the species, a number similar to that typical of various social bird species, such as domestic fowl, pheasants, or rock doves. Of seven vocalizations studied by W. J. Smith and

Fig. 10. Black-tailed prairie dog, adult jump-yip display.

others (1977), each such signal provides information that identifies its user, correlates with the probable occurrence of a set of activities, and provides information on at least five current or forthcoming behaviors. Tail movements, such as flicking, are important visual signals, as are general postural variations.

In contrast to the general contact calls of the other U.S. prairie dogs, the black-tailed has evolved a two-syllable yipping note. It is accompanied by a quick upward and backward jump as the mouth is opened wide. The quickness and rather comical aspects of this performance have earned it the name "jump-yip" display, although it is also widely known as the "all-clear" call (Fig. 10). The display takes less than a second to perform and is usually done as the animal is standing on its burrow mound but may occasionally be uttered from within the burrow. It quite regularly occurs after attacks or threats by predators are over, suggesting an "all-clear" function. It also may occur during territorial interactions and in other less-specific situations. In most cases the animal is less likely to flee following a jump-yip than immediately before, but almost any activity other than attack may occur.

Like the other prairie dogs and other less social rodents, an important social signal of the black-tailed is nose-to-nose "kissing" behavior between close relatives, such as mother and offspring or mated adults (Fig. 11). This brief

Fig. 11. Black-tailed prairie dog, adult and young "kissing" as social greeting. After a photo in Hoogland 1994.

encounter probably serves as both a tactile and especially an olfactory signal between individuals that are well known to each other and are part of their own coterie. Mutual grooming – involving the removal of fleas, ticks, and lice from another individual in the coterie – is also an important aspect of social behavior that probably has significant survival implications.

Probably the strongest social bonds in a prairie dog colony are those formed between a nursing female and her dependent offspring. The typical litter size is of three to five young, but as many as eight have been reported, corresponding to the number of teats on the female. In John Hoogland's (1994) study, he found that about half of the breeding females observed were able to rear young to first emergence from the burrow, with infanticide accounting for most of the losses. Typically, lactating females kill the young of close kin living in the same coterie, usually killing all the pups in a litter. Some females specialize in such killing, but in doing so they place their own young at risk. Possible benefits of this remarkable behavior include the removal of future competitors, increased personal sustenance and foraging areas, additional help from victimized mothers who may then become helpers, and the reduced probability of losing one's own litter to infanticide. Females immigrating into a coterie are infanticidal, and mothers sometimes even abandon their own young soon after

Fig. 12. Black-tailed prairie dog, female nursing juvenile. After a photo in Graves 2001.

parturition, allowing them to be cannibalized. Even more oddly, an invading male sometimes kills younger half-siblings, namely, individuals that were sired by his own father. Probably most deaths of young within a coterie come from such infanticide by lactating females, which takes the lives of almost half of all the young that are born. Such predatory females often consume the young that they have just killed (Hoogland 1994).

The young weigh about half an ounce at birth, which occurs after thirty-four to thirty-five days of gestation. Like other rodents the young are born hairless and with their eyes tightly shut. They gain about 40 percent of additional weight in their first week postpartum. By three weeks of age hair has begun to develop over their bodies, and at five weeks their eyes open. At about six weeks the young weigh on average about 5.2 ounces, and the mean litter size at this usual time of first emergence from the burrow is about 3.1 pups (Hoogland 1994). By this stage the pups are shifting from a diet of milk to one of solid foods. Some nursing continues above ground for a time after the young emerge (Fig. 12), but care is taken by each female parent that its young not stray too near another mother, as infanticide may account for the partial or total loss of about 40 percent of all litters.

However, the greatest mortality source for prairie dogs by far is humans, who for more than a century have regarded prairie dogs as satanic creatures fit only to be poisoned, shot, drowned, or trapped, whenever and wherever they are found. This threat will be discussed in greater detail later.

Sections 1 to 4 of this act shall be known as the Prairie Protection Act. For purposes of the Prairie Protection Act, destructive rodent pests means one or more rodents, including but not limited to prairie dogs, ground squirrels, pocket gophers, jackrabbits and rats. . . . The county board may purchase material and equipment and employ one or more suitable persons to eradicate destructive rodents within the county. – *Legislative Bill 353, 98th Nebraska Legislature, 2003*

4

Ferrets, Badgers, Bobcats, and Coyotes

Coping with Dangerous Neighbors

The Coyote said, "I will paint with blood. There shall be blood in the world; and people shall be born there, having blood. There shall be birds born who shall have blood. Everything—all things shall have blood that are to be created in this world." – *Maidu myth*

It is Coyote who wanders naked in the desert and leaves his skin on the highway, allowing us to believe he is dead. – *Terry Tempest Williams,* Red: Passion and Patience in the Desert *(2001)*

The vertebrate predators of prairie dogs are legion. As John Hoogland (1994) has summarized, they often come in the form of mammalian predators, of which the most important have historically been black-footed ferrets, badgers, bobcats, coyotes, weasels, and especially humans. Less important mammalian predators are likely to include red and gray foxes, mountain lions, and (historically) grizzly bears. Then there are the larger winged raptors, which will be discussed later. Last are diamondback and prairie rattlesnakes and the bullsnake, which – like black-footed ferrets – are able to enter burrows easily and attack without warning. As an unpleasant afterthought, prairie dogs also sometimes commit infanticide (discussed in Chapter 3), so young animals are not even entirely safe from adults of their own species.

Thus prairie dogs are potentially under attack at all times – from above, from all sides, and even perhaps from below, once a snake or ferret enters a burrow. As a result prairie dogs spend most of their daylight hours scanning for predators. When they see a predator approaching from a distance, they are likely to crouch at the burrow's entrance while uttering repeated alarm calls to alert all nearby animals (Figs. 13 and 14). They probably also do not sleep very soundly at night, at least when ferrets and snakes may be about. Only very occasionally can they afford to relax enough to stretch out and sunbathe for a time beside their burrow (Fig. 13). Therefore a prairie dog that has lived as long as four years would have to count itself very lucky.

Fig. 13. Black-tailed prairie dog, antipredator crouching (above) and sunbathing. After various sources.

Fig. 14. Black-tailed prairie dog, adult alarm note. After a photo in Hoogland 1994.

Of all the predators of prairie dogs, the black-footed ferret is easily the most famous (Figs. 15 and 16). Theodore Roosevelt (1900), in describing his ranching experiences in western North Dakota, referred to the black-footed ferret as the "arch enemy" of prairie dogs. He also described how one of his ranch hands had seen a ferret take the throat of a pronghorn fawn, "sucking its blood with hideous greediness." One wonders whether he had actually seen a long-tailed weasel performing this activity, as black-footed ferrets feed almost exclusively on prairie dogs and they were already becoming quite rare in North Dakota by the time Roosevelt reported these observations. Long-tailed weasels are considerably smaller than black-footed ferrets but are equally fearless and are also known to attack prey much larger than themselves.

The story of the black-footed ferret, one of America's rarest and most charismatic animals, has been told many times. Its original range coincided closely with that of the black-tailed prairie dog and all other prairie dogs except for the Utah and Mexican species. In one food-habits study, prairie dogs constituted 82 percent of the identified remains found in fifty-one ferret scats. The other nineteen scats had only mouse remains. In two other studies the incidence of prairie dog remains were 87 percent of eighty-six scats in Wyoming, and 91 percent of eighty-two South Dakota scats. Other prey that have been recorded include ground squirrels, pocket gophers, rabbits, microtine voles, birds, and even insects (U.S. Fish & Wildlife Service 1988).

Ferrets are largely nocturnal and are small enough to make their way through prairie dog tunnels. They live in the tunnels and may remain below ground for weeks at a time during winter conditions. Ferrets also do some digging where they find dirt plugs in a tunnel. A substantial percentage of black-tailed prairie dog burrows have such plugs, and some occur in the burrows of white-tailed prairie dogs. Ferrets are not rapid diggers as compared with badgers and may require two or three hours to finish a dig, perhaps ample time for a prairie dog to escape if not trapped in a tunnel with a dead end.

Ferrets have high metabolic rates. It has been estimated that about 150 pounds of prairie dogs might be needed to sustain a single ferret for a year. For an adult female rearing an average-sized litter of three or four young, this would translate to about 109 adult prairie dogs annually, or roughly one every three days. For a single adult ferret weighing 1.5 to 2.2 pounds, a full-sized black-tailed prairie dog might last six or seven days. Ferrets have been known to cache prairie dog carcasses for later consumption, and they may also resort to eating prairie dog carrion.

Females receive no help from males while caring for their young; the animals are essentially solitary except during the short spring breeding period. After the

Fig. 15. Head of a black-footed ferret. After various sources.

Fig. 16. Adult black-footed ferret. Sketch above shows a ferret in a prairie dog burrow (after a drawing in Miller, Reading, and Forrest 1996).

young ferrets have been weaned, they move varying distances beyond the natal area. Females tend to remain near their natal homes, generally within 1,000 yards, whereas males may disperse up to several miles. However, once they are adults (at one year old), ferrets develop a high fidelity to their foraging areas, often remaining within a single prairie dog colony (Miller, Reading, and Forrest 1996). Ferret densities on white-tailed prairie dog habitat in Wyoming were estimated at about one adult per 99 to 148 acres of prairie dog colony. Female ferrets with litters have not been found in white-tailed prairie dog colonies smaller than 121 acres, and single adults have not been observed for at least a year using a colony smaller than 31 acres (U.S. Fish & Wildlife Service 1988). Ferrets have been reintroduced into large prairie dog colonies in several U.S. states (Montana, Colorado, Utah, Wyoming, South Dakota, and Arizona) and recently (2001 and 2002) into northern Mexico. As of 2002 there were 264 ferrets in South Dakota, 15 in Wyoming, 38 in Colorado and Utah, and 11 in Montana.

Although ferocious animals, even black-footed ferrets have several predatory enemies, including golden eagles, great horned owls, coyotes, and badgers. Of sixteen confirmed deaths of captive-raised animals that had been released into the wild, twelve were caused by coyotes (Miller, Reading, and Forrest 1996). Evidently a serious mortality factor, canine distemper was the primary cause of the collapse of the last known wild population of ferrets in Wyoming. Ferrets are also vulnerable to sylvatic plague and are also sensitive to rabies, tularemia, and human influenza. Like prairie dogs, ferrets do not have long life spans in the wild, with estimated annual mortality rates of adults estimated at about 60 percent, and few animals are believed to live more than four or five years.

Through unrelenting prairie dog poisoning campaigns by state, federal, and private parties, the food base of the black-footed ferret eventually collapsed. As a result it was one of the first mammals to make its way onto the list of nationally endangered species after the Endangered Species Preservation Act was passed in 1966. This opened the door to providing moneys via the U.S. Fish & Wildlife Service for the recovery of black-tailed ferrets ($1.5 million was spent on the species in 1991, for example). At the same time, through their rodent control programs, the Departments of Interior and Agriculture continued to provide massive amounts of money and poison for eliminating prairie dogs, the very species the black-footed ferret needed for its own survival.

If such a mess can be made of efforts to save an attractive creature such as the black-footed ferret in a country as well organized as the United States, prospects for conservation in other parts of the world are indeed bleak. – R. M. May (quoted in Miller, Reading, and Forrest 1996)

Fig. 17. Adult badger.

AMERICAN BADGER

In many areas the American badger (Fig. 17) seems to be one of the most important present-day predators of prairie dogs, as they are frequently found near prairie dog towns. However, a summary of diverse dietary data reported by Long and Killingley (1983) indicated that only in northern Mexico, where Mexican prairie dog remains made up 32.5 percent of forty scat remains in one study, has the prairie dog been reported as a major part of the species' present diet. It is likely that in earlier days, when prairie dogs were far more abundant, this ratio was quite different. In Mexico, badgers were found to live among and feed on prairie dogs, especially in November, when the badgers were forming pairs and thus associating. It would seem logical that they should hunt in pairs for maximum efficiency in hunting animals like prairie dogs, but the species is essentially solitary, except during the brief autumn mating season.

In any case, the primary prey of North American badgers is always rodents, the specific species probably varying greatly by region and relative prey abundance.

Kangaroo rats, mice, and ground squirrels are all taken locally, the small rodents more frequently than the large ones. The badger's top running speed is only about 6 miles per hour, or perhaps 12 miles per hour for short distances, but this is far too slow to catch a running prairie dog unless it becomes trapped in a dead-end burrow. The badger digs so rapidly that it can probably capture a prairie dog before the latter has been able to remove a protective plug within its burrow in order to escape out another opening. Badgers have large home ranges, with males tending to move about more than females, and young animals more than adults. Among several studies, home ranges of adult males have ranged from about 600 to 4,000 acres, and those of females from about 400 to 600 acres. With such ranges, several different prairie dog colonies could be visited by a single animal, especially in the case of immature animals or adult males.

There have been several reports of badgers and coyotes forming predatory pairs, one of the first being noted by E. T. Seton (1929). The badger digs out burrowing rodents, and the quick-footed coyote catches any that happen to try to escape above ground. In one case a coyote was observed to be followed by a badger, with the badger trying its best to keep up with the coyote, suggesting that the benefits of team hunting may not always be simply to the coyote's benefit. However, badgers are not always on the best terms with large canids such as coyotes and wolves, and in at least two cases a badger was observed to be attacked when outnumbered, and in one case was killed. Despite the badger's value in controlling rodents, it has long been persecuted by both fur trappers and government predator control agents. In California, U.S. Fish & Wildlife animal control agents killed ten times as many badgers (4,086) between 1965 and 1976 than did fur trappers, even though agricultural livestock, such as cattle, sheep, and even poultry, never turn up in badger dietary studies. The daily dietary needs of a captive 18-pound badger has been estimated as about a fifth of a cottontail rabbit or an entire 3-ounce ground squirrel. Wild-living badgers certainly need more energy, but about two pocket gophers (which weigh about 7 ounces, or about 20 percent of an adult prairie dog) were believed adequate to satisfy the daily needs of an adult badger during summer months (Long and Killingley 1983). Thus an adult black-tailed prairie dog would provide a two-day food supply for an adult badger.

Badgers have received very little protection by individual states, despite their contributions to small rodent control. No protection at all exists for them in many states, including Montana and Nebraska. Only seasonal protection is provided in some other states, including Wyoming, Oklahoma, and New Mexico, which allow seasonal trapping by paid permit. Only in a few states, including Michigan, Wisconsin, and Ohio, are badgers given complete year-round protection. As of the early 1990s an estimated 42,000 badger pelts were being harvested annually. The species' long, silvery fur is used in paintbrushes, shaving brushes, decorative rugs, and trim for coats. Badger pelts brought about $10 each as of 2002 but have ranged as low as $3 each in the past fifteen years.

"Every ignorant person will kill a badger." Most of the badgers killed are killed – not ignorantly – by government trappers after coyotes. Badgers come readily to scents, poisons, traps placed for coyotes. Their holes are a nuisance to range riders, but a benefit to the soil. Badgers control rodents. Their fur is an economic asset. Their existence delights all people cultivated enough to know that the pursuit of happiness is not limited to killing out of doors and making money indoors. – *J. Frank Dobie*, The Voice of the Coyote *(1949)*

BOBCAT

John Hoogland (1994) reported that he and his team of researchers saw more cases of prairie dog predation by bobcats than by any other predator; the species was implicated in nine of twenty-six total observed predation events. But the bobcat (Fig. 18) is primarily nocturnal and is mainly active when prairie dogs are not likely to be above ground. Nevertheless, bobcats are also highly opportunistic, taking whatever prey might be locally available, wherever and whenever it can be found, from mammals as small as mice to as large as deer. Bobcats also take other vertebrates they can capture and at times may even consume insects. Compared with badgers, bobcats have very large home ranges, with males wandering more than females and covering up to as much as 75 square miles. Since other observers have not mentioned bobcats as a notable threat to prairie dogs, it seems likely that the cases of prairie dog predation observed by Hoogland may have represented a few individual animals that had learned how to specialize in prairie dogs, as mammalian predators sometimes do.

In Stanley Young's now rather old (1958) but comprehensive summary of bobcat foods (which involved 3,990 stomach contents), prairie dog remains comprised only twenty-five items, or 0.6 percent of the total. Rodent and rabbit remains collectively totaled 51.3 percent of the items; livestock, 13.6 percent; and deer or pronghorn, 2.5 percent. Among birds, wild species represented 14.7 percent and poultry only 3.3 percent. Trap bait remains comprised 6.8 percent, and small quantities of carrion, fish, frogs, reptiles, insects, and vegetation made up the remainder. Among 3,538 bobcat stomachs obtained from thirty mostly western states, rodents predominated and were found in 46 percent of the

Fig. 18. Adult bobcat.

stomachs. Rabbits occurred in 45 percent of the stomachs; deer, in 5 percent; and livestock, in 2 percent. Other mammal remains were found in 2 percent of the stomachs, and smaller numbers of game birds, nongame birds, poultry reptiles, and invertebrates were also present. Some prairie dog remains were found in bobcat stomachs obtained from New Mexico, Colorado, Utah, and Wyoming. One of Young's contacts believed that prairie dogs were a favorite food of bobcats in the Custer National Forest area of Montana, and other contacts in New Mexico, Colorado, and Arizona had observed bobcats chasing or catching prairie dogs. Of the livestock taken, sheep and goats were the most common prey (11.8 percent of the 3,990 food items found in one study; present in less than 2 percent of the stomachs in the other), but there is no way to sort out the sheep and goats that were actually killed by bobcats from those that were found dead and consumed as carrion.

Bobcats range over the entire North American region occupied by prairie dogs but, in general, occur in low densities of from as few as one to as many as seventy individuals per 100 square miles. However, the animals' large individual home ranges may allow them to visit a considerable number of prairie dog towns periodically, and without unduly sensitizing the prairie dogs to their presence or hunting techniques. Like other cats they are primarily stalk-and-spring hunters, and the open habitats favored by black-tailed prairie dogs would seem to put bobcats at a strong disadvantage while stalking.

Substantial numbers of bobcats are still trapped for their skins. Bobcat pelts brought about forty-five each as of 2002 but their prices have ranged from about $28 to $70 in the past fifteen years.

It is certain this carnivore consumes a wide variety of wild prey. This, coupled with the new and satisfying provender in the young of the flocks and herds of the stockman and the poultry farmer over much of its natural range causes it to be a close second to the coyote in its adaptability to any modified habitat. – *Stanley Young (1958)*

COYOTES

Of all the mammalian predators of prairie dogs, probably more has been written about predation by coyotes (Fig. 19) than any other species, perhaps because it can be observed fairly easily. It may be more accurate to say that predation efforts by coyotes can be observed frequently, for although coyotes can often be seen loitering suspiciously around dog towns, it is most unusual to see them make a successful capture. Of the prey animals reported from stomach analysis of some thirty thousand coyotes killed by predator control agents between 1919 and 1923 (see Table 4), prairie dog remains constituted only 584 prey records out of a total of 36,989 items found, or 1.6 percent of the total. These data were obtained at a time when prairie dogs were still abundant in the West. The data also were based on specimens obtained by government trappers dealing with supposed problem coyotes on livestock ranges, so the incidence of livestock remains relative to that of wild game may be somewhat greater than would be true in a random sample.

Rodents generally are the main food of coyotes, not sheep, cattle, or even domestic poultry, as many people raising cattle or sheep would argue. In a broad survey of coyote foods using data from across the species' original range, P. S. Gipson (1974) reported that rabbits and rodents were present in 77.3 percent of the species' food samples; carrion, in 25.5 percent; livestock, in 21.9 percent; wild birds, in 15 percent; deer, in 7.9 percent; fruits, in 6.7 percent; and poultry, in 4.8 percent. Most of the large wild mammals and domestic livestock taken by coyotes are sick, old, or young animals. Likewise, most lambs eaten in spring have already died from starvation, disease, abandonment, tail docking, or infection. In Adolph Murie's classic study of coyotes in Yellowstone (1940), he found that 70.3 percent of the animal's diet would be classified as "beneficial"

Fig. 19. Adult coyote.

to humans, and only 11.5 percent as "harmful." Of the many studies that have been done on coyote ecology, little evidence has accumulated to suggest that coyotes may be a primary limiting factor in the populations of either domestic livestock or wild ungulates (see Table 4).

Like bobcats, coyotes are widely ranging predators but are diurnal rather than nocturnal and so tend to encounter a different array of potential prey. They are also animals that rely more on their ability to perform prolonged chases rather than on the patient stalking and pouncing behavior of bobcats. They can run at speeds of at least 30 miles per hour for extended periods and will eat virtually anything that offers nourishment, even including discarded leather boots and garbage. Coyotes are most likely to take large mammalian prey in winter as well as considerable amounts of fruit locally during autumn.

Like wolves, coyotes are nearly always on the move and tend to hunt mostly near dusk, with a secondary peak of activity at dawn. Coyotes have very large home ranges that, in various studies, have been found to range from about 3 to 25 square miles, with males sometimes having larger home ranges than females. A coyote may often range over about 2 miles daily, and over long periods movements of as much as 100 miles are common.

Coyotes are less social than wolves; the expression "lone coyote" has some

Table 4. Coyote foods based on stomach analyses

	STUDY 1 [a] FREQUENCY(%)	STUDY 2 [b] FREQUENCY(%)	STUDY 3 [c] VOLUME(%)
SAMPLE SIZE	ca. 30,000	ca. 50,000	8,263
FOODS			
Rabbits	26.4	24.3	33.25
Bait	17.6	23.3	12.0
Livestock			
Sheep or goats	23.1	18.2	13.0
Other livestock	9.1	5.2	0.7
Poultry	4.3	5.4	0.75
Total livestock	36.5	28.8	14.45
Rodents			
Mice and rats	4.75	5.2	
Ground squirrels	3.8	3.5	
Prairie dogs	1.9	1.0	
Other rodents	0.4	0.8	
Total rodents	10.85	10.5	17.5
Carrion	7.8	7.7	13.2
Wild ungulates	1.5	2.7	3.6
Other mammals			
Vegetable matter	7.4	6.2	1.75
Wild Birds			
Grouse/quail	4.2	5.1	
Other birds	3.0	2.2	
Total birds	7.2	7.3	1.25
Miscellaneous	2.7	1.3	1.3

[a] After Young and Jackson 1951. Frequency figures do not add to 100%, as some stomachs were empty and others probably contained more than one item.

[b] After Sperry, 1941.

[c] Out of a total of 14,829 stomachs, excluding empty stomachs, those with debris only, and those from un-weaned pups (Sperry 1941).

truth to it. Some areas may support "packs" of coyotes at times, perhaps depending on the abundance and type of their primary prey, but the coyote is only slightly more social than are foxes, and pair and family bonds are the primary social unit. It is thought that the same individuals pair from year to year but do not necessarily mate for life. In the wild, few coyotes probably live more than six to eight years, and sexual maturity occurs in the first year. The same den is often used in successive years, and in a few instances two litters have been found in the same den during a single year.

Coyotes do deposit scent marks, but it is not certain that these serve as territorial markers. For short-distance signaling, coyotes use postures, gestures, and facial expressions that closely resemble those of domestic dogs. Much of the coyote's long-distance social communication is achieved by vocalizations; at least eleven distinct vocal signals have been detected. Such signals are invisible yet travel fast and far, an important advantage for an animal that cannot afford to be seen and is regularly on the move. Densities of about one coyote per 1 to 2 square miles are probably typical for large areas. They are relatively rare in areas occupied by wolves, which often kill coyotes, and mountain lions will also both kill and consume coyotes. In turn, coyotes will not tolerate bobcats or foxes and are also a serious threat to swift foxes, which they regularly kill.

Whether or not coyotes represent a significant threat to prairie dogs, swift foxes, or other High Plains wildlife, they are as much a part of the shortgrass ecosystem as the pronghorn, the bison, and the prairie dog. It is an exhilarating experience to hear coyotes yipping in the moonlight of a summer night when camping in the West. The coyote offers solace because – although most of our western wildlife has either disappeared or become so rare that there is little hope of experiencing it – this animal remains an enigma. It persists as a singularly beautiful creature, at once elusive, disdainful, distrustful, and defiant of humans, always wholly in its element, whether it be the arid deserts of Mexico and the Southwest or the newly colonized dense forests of eastern North America. It is almost certainly this remote, untamable aspect of the coyote that has made so many people despise it; it is easier to try destroy a creature that cannot be controlled than it is to try understand it.

Stanley Young, who wrote *The Clever Coyote* (1951) and directed the activities of hundreds of federal government predator control agents sent out to kill coyotes over several decades during the mid-1900s, confidently believed that predator control was imperative and that "we find the government hunters as a group capable to coping in a most satisfactory manner with any vexing problem that may arise" (p. 179). He also offered a revealing attitude toward predators such as coyotes by suggesting that "probably most of the mass killing by any predatory animal indicates that it is in the height of its vigor and thus kills through sheer physical exuberance. Certain of the Eskimo and Indian tribes so history shows us have had at times the same lust for killing when occasion offered, and it is not wholly absent from men higher up the scale today" (p. 225).

As of 2002, a dead coyote brought only about $15 for a prime pelt. A living coyote is worth far more than the sum of all its nonliving parts and, indeed, cannot be assigned a monetary value. In mythology, it is a remote but direct spiritual descendent of the Aztec god Tezcatlipoca, who could transform himself into a coyote at will. The coyote is also God's Dog of the Navahos, somewhat related to the Skinwalker, who could transform himself from a human into a coyote and take appropriate vengeance on his enemies. He is also known by the Navajo as Ma'ii, the one who is never to be taken for granted. In some northwestern tribes such as the Maidu, the Coyote was the younger brother of Kodoysanpe the Earth Namer, a major Creator figure. The coyote was also the Trickster of many arid plains and desert tribes. In this incarnation Coyote represented a sort of Janus-like figure representing both good and evil, but mostly causing strife and hardship, and a force able to control the lives and destinies of mere humans. According to the Shoshonean Utes, in Early Times the destiny of the world was once controlled by two brothers. They were Wolf, the older one, and Coyote, the younger one. These brothers made many of the important decisions about our world with which humans have to contend today. Given the historically long and ecologically close association of Native Americans with the coyote, it is probably safe to say that Coyote has been given credit for as many miraculous or mysterious events as any biblical figure.

Once Wolf and Coyote debated the fate of humans. Coyote suggested that humans should be provided with many seeds and fruits, so that they need never go hungry. Wolf objected, saying that if such were the case humans would become lazy and irritable. Coyote then suggested the humans should have their daily thirsts quenched with honeydew from heaven. This too Wolf denied. When Coyote approached Wolf a third time he suggested that after a man dies he should remain dead only for one night, and then return to life the next morning. But Wolf insisted that death be permanent. Some time later Coyote saw Wolf's son at play, and killed him. When Wolf came looking for his son, Coyote told Wolf that he was dead, and added that, "You made the law that the dead shall never return. I am glad that you shall be the first to suffer." – *Ute myth, adapted from Powell 1881*

5

Free-loaders and Hangers-on

The Rewards and Dangers of City Life

The [prairie dog's] toe nails [are] long, they have fine fur and the longer hairs is gray, it is Said that a kind of Lizard also a Snake reside with these animals (did not find this correct). – *William Clark, September 7, 1804, near the mouth of the Niobrara River, Nebraska*

From earliest times, it has been known that prairie dogs do not live alone in their towns but, rather, serve as compliant hosts to many uninvited species. These variably brief to long-term visitors find that the newly sprouting greenery associated with the colony offers a source of fresh, nutritious food; that panoramic visibility afforded by the dogs' lookout mounds suits their own sentinel purposes; and that the dogs' burrows offer a safe refuge from danger or a cool escape from summer's heat and winter's frigid blasts. Dozens of other reasons – some more obvious than others – also explain why more than two hundred species of animals have been reported in and around prairie dog colonies. Some of these species visit the dog towns much more frequently and consistently than do others.

Of all these visitors, one of the species most consistently and famously associated with prairie dogs is the burrowing owl, a small, diurnal Western Hemisphere owl that is most often found where prairie dog burrows occur, or at least where similar-sized burrows made by large rodents or other burrowing animals occur. Only in a few areas, such as Florida, where no larger rodents are present, does the bird need to sometimes dig its own burrows. Its geographic distribution in western North America (Map 6) still centers on the collective historic ranges of prairie dogs but, like that of prairie dogs, has retracted greatly in recent decades, especially along its eastern edges where prairie dogs have been eliminated.

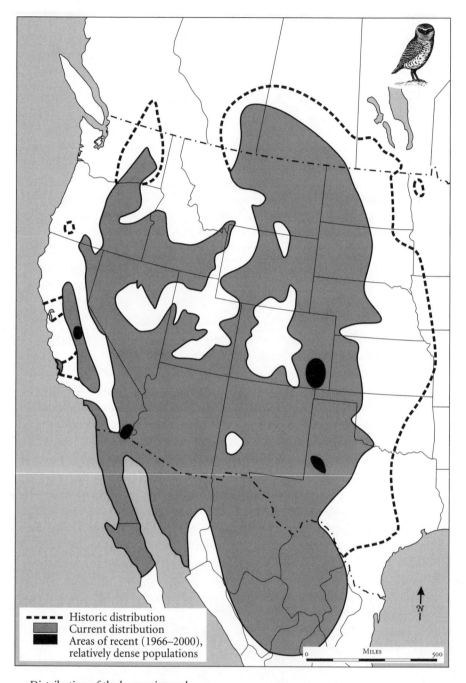

Historic distribution
Current distribution
Areas of recent (1966–2000), relatively dense populations

Miles
0   500

Distribution of the burrowing owl

The owl moves at night when people are asleep. The medicine-man gets his power through dreams at night and believes that his dream is clear, like an owl's sight . . . so he promises he will never harm the owl. – *J. R. Walker,* The Sun Dance and Other Ceremonies of the Oglala Division of the Teton Dakota *(1917)*

Burrowing owls, like Westerners generally, tend to be obstinately independent and generally refuse to obey the rules of owl decorum set out for them. First, they do not – as stated in the Walker quote above – move about at night in the usual manner of owls but instead are active mainly during bright daylight. Second, unlike the larger owls of the grasslands, burrowing owls are surprisingly insectivorous, at least during summer months when large, slow-moving insects, such as dung beetles and ground beetles, are abundant around dog towns and are easily captured. Third, burrowing owls seem to lack the acute binaural hearing and precise sound-source localization abilities that characterize owls as a group and instead appear to rely on their keen daytime vision for finding prey. Given these traits, one might think that the burrowing owl is only a second-rate member of the Strigidae – that is, until one actually sees a burrowing owl for the first time (Fig. 20).

Few birds will bring bird watchers screeching their cars to a halt faster than will a burrowing owl perched quietly on a fence post or peering quizzically out of a prairie dog hole. There is something in the animal's intense, yellowish green eyes that demands to be watched with equal intensity, and its comical, not-quite-erect stance might remind one of a spindly legged, feather-clad leprechaun still trying to recover from last night's hangover.

Often, after a few slight horizontal or vertical head movements, as if the bird were trying to shake its head clear of a foggy memory but which probably allow it to get a better distance-estimating fix on the intruder into its personal space, the owl will probably either duck back into its hole or, if perched on a post, perhaps take silent flight over the prairie dog town and land near its burrow entrance. When cornered in its burrow, a owl will utter a rattling call much like the sound of a rattlesnake's rattle and may also spread and tilt its wings vertically, apparently to make itself appear larger and more dangerous than it actually is.

Although a burrowing owl is perfectly content to take over a prairie dog burrow without making major structural changes or other renovations, it is likely to gather nearby pieces of dried bison or cattle dung, break them into small pieces, and line the entrance area in front of its burrow with these bits of debris. Such markings help delineate an active burrowing owl burrow, as do the

Fig. 20. Adult burrowing owl, defensive threat.

dried owl pellets that also appear there, which are usually rich in chitinous insect fragments, such as the undigested exoskeletal remnants of grasshoppers and scarab and carabid beetles. The function of the scattered ungulate droppings is uncertain, but they may help mask the odors of an active owl burrow, for burrowing owls are just as vulnerable to badgers as are prairie dogs, and badgers seem to find young burrowing owls just as tasty as baby prairie dogs. Gregory Green and Robert Anthony (1989) found that, next to desertion, badgers were the most significant source of nesting mortality for burrowing owls but that nests lined with livestock dung were significantly less prone to predation than unlined nests. It has also been suggested that the presence of dung helps attract dung beetles to the nest, where they can be easily captured by the owls.

The burrowing owl is the only strongly migratory owl of the northern plains, presumably because its insect foods become scarce in fall. Then it turns increasingly to small rodents, such as pocket mice for food, though I also once saw an adult burrowing owl dragging a dead, thirteen-lined ground squirrel over the ground toward its burrow. I suspect it was simply cashing in on a road kill, as the ground squirrel's body weight (5 to 9 ounces) would be somewhat heavier than the owl itself (about 5 ounces), and it is hard to imagine any burrowing owl quite so ambitious as to take on a healthy ground squirrel. Although small

mammals and birds make up a small percentage of the summer diet when measured by their sheer number, the much larger body mass of mammals relative to
insects tends to make mammals account for the majority of the biomass of all
foods taken. Furthermore, some owl activity occurs at night, especially during
bright moonlit nights, and at that time such nocturnal mammals as voles and
heteromyid mice, rather than insects, would be the more likely targets. Frogs,
toads, lizards, snakes, and even turtles have also been reported as burrowing owl
foods (Haug, Millsap, and Martell 1993). Ray Poulin (2003) determined that in
Saskatchewan the primary owl foods are deer mice and meadow voles and that
the only year the owls showed a population increase was following a peak in
vole populations.

Besides using prairie dog burrows, burrowing owls have also occasionally
adapted to living in burrows made by ground squirrels, marmots, woodchucks,
badgers, foxes (swift, kit, and red), coyotes, skunks, nine-banded armadillos,
kangaroo rats, and tortoises. They may also dig their own burrows in the absence
of available housing already provided by burrowing mammals. They have even
been found using natural rock cavities and humanmade artificial burrows where
natural excavations are not possible. Yet, on the Great Plains, is it primarily the
black-tailed prairie dog that offers the owls housing. An unpublished study
by Shawn Conrad, Jennifer Dose, and Dan Svingen estimated that 95 percent
of all the burrowing owls nesting on the Comanche, Cimarron, Rita Blanca,
and Kiowa national grasslands, all located on the southern Great Plains, were
directly associated with prairie dog towns. Only 3 of 114 prairie dog towns
supporting burrowing owls were inactive, indicating the importance of dog
activity in making good owl breeding sites. The 3 inactive towns supporting
burrowing owls had apparently been inactive for only a year. Likewise, among
543 burrowing owl nests found in the Oklahoma Panhandle, 66 percent were
found within black-tailed prairie dog colonies (Butts and Lewis 1982). Besides
being able to exploit the inactive dog burrows, the owls favor the combination
of low shrub coverage, short vegetation, and high percentage of bare ground
that is typically present in prairie dog colonies. Where the ground-level vision
is variously obstructed, the birds always find suitable observation posts to use
nearby, such as fence posts, boulders, or utility poles.

Within prairie dog towns, burrowing owls favor larger and active colonies;
typically a town that has been abandoned by prairie dogs will also be abandoned
by burrowing owls within three years, largely because of the encroachment of
dense vegetation but also because of the gradual deterioration of the burrows
themselves. In active towns the owls often choose nests near the periphery of
the colony, which may have more insects, more available nearby perches, and
a proximity to foraging areas. Burrowing owls may also favor sites offering

several "satellite" burrows, used by both adults and young, perhaps to avoid nest parasites. As many as five such satellite burrows may be used by a single owl family (Dechant et al. 1999).

For the Great Plains as a whole, John Sidle and others (2001) reported that burrowing owl occupancy of active prairie dog colonies was lower in the northern Great Plains national grasslands (59 percent occupancy) than in those of the southern Great Plains (93 percent). Regardless of geographic location, burrowing owls were found to use active dog towns as a rate 6.3 times greater than at inactive ones, further indicating the need to maintain active prairie dog colonies if maintaining breeding habitat for burrowing owls is desired. The presence of live prairie dogs means that a ready supply of newly abandoned but structurally intact burrows will be available to the owls, that the owls may exploit the efficient warning system of the prairie dogs when predators approach, and that the prairie dogs' dung may attract dung beetles and other food sources. No evidence exists, however, that prairie dogs directly benefit from the presence of burrowing owls.

A recent survey (1997) of estimated breeding populations by Paul James and Rick Espie suggested that in 1992 there may have been 20,000 to 200,000 breeding pairs in the United States plus 2,000 to 20,000 pairs in Canada. In the Great Plains, states or provinces having population estimates of 1,000 to 10,000 pairs were Colorado, New Mexico, Wyoming, and Saskatchewan, and Texas estimated a population in excess of 10,000 pairs. States with 100 to 1,000 estimated pairs were Kansas, Montana, Nebraska, North Dakota, Oklahoma, and South Dakota. Manitoba had 10 to 100 pairs. The species is nationally endangered in Canada, is state threatened in Colorado, and is considered a species of concern in Montana, Wyoming, and Oklahoma. It is unlisted in the other Great Plains states but is considered vulnerable by the state natural heritage divisions of Kansas and Nebraska. In western Nebraska one population study indicated that a 58 percent population decline occurred between 1990 and 1996 (Desmond and Eskridge 2000). Breeding bird surveys have indicated a generally declining population trend in the northern half of the Great Plains.

The Mexican population of burrowing owls is unknown, but the species is classified as threatened. It includes both breeding populations in prairie dog towns of northern Mexico and local wintering populations south to the Isthmus and beyond. Historically, breeding may have extended locally to the Isthmus and the Yucatan Peninsula. An excellent conservation plan and status assessment for the species has recently been published (Klute et al. 2003), concluding that the preservation of native grasslands and burrowing mammal populations will be critical to the long-term conservation of this hallmark grassland species.

Fig. 21. Adult swift fox. After various sources.

SWIFT FOX

There is a remarkable small fox which associate in large communities and burrow in the prairies something like the small wolf [coyote], but we have not as yet been able to obtain one of them; they are extremely watchful and take refuge in their burrows which are very deep. – *Meriwether Lewis, July 6, 1805*

Like the burrowing owl, the swift fox (Fig. 21) pledged its troth to the prairie dog ecosystem long, long ago and is now suffering the corresponding consequences. Its range and populations have been gradually slipping away, almost as rapidly and silently as Lewis Carroll's Cheshire Cat could slowly disappear from view. Only a small remnant population of swift foxes remains, whose area stretches in a narrow north-south band, resembling the crooked grin of the Cheshire Cat, from the boundary area of Montana and Canada south to eastern New Mexico and western Texas (Map 7). Along this High Plains zone the swift fox is now confined largely to shortgrass areas under state or federal protection, such as the national grasslands. Its almost identical relative the kit fox is apparently surviving better and has a much larger remaining range. The ranges of these two very closely related foxes come into geographic contact in western Texas and southeastern New Mexico. They do occasionally hybridize in these areas of contact but nevertheless have traditionally been regarded as representing distinct species. However, studies in southwestern New Mexico suggest a high level of interbreeding and thus a questionable degree of species discrimination. Recent ongoing surveys in New Mexico reported the presence of swift fox scats in 71 percent of ninety-nine transects east of the Pecos River, suggesting a better than expected New Mexican population.

The swift fox is well named; both its English vernacular name and its Latin specific name (*velox*) describe its rapidity. It is tiny and lightning fast and can hit a top speed of about 37 miles per hour from a standing start and come to a standing stop just as rapidly. Like the very closely related kit fox, it has enormous ears, a telltale sign that it evolved in a warm, desertlike environment, where the dangers of frostbite are nil and where, because sound does not travel well in dry air, each minute pulse of sound energy must be extracted from the environment and identified as to source and significance if the animal is to locate both its food and the presence of possible enemies. It is small enough to perhaps be able to duck into the burrow of a prairie dog when threatened, and its historic range closely corresponded with that of the black-tailed prairie dog.

Swift foxes generally dig their own dens, and like those of prairie dogs, their abandoned dens may attract other species, including prairie dogs, burrowing owls, prairie rattlesnakes, deer mice, striped skunks, and Great Plains toads. Invertebrates also use their burrows, especially beetles, spiders, and fleas. Swift foxes sometimes also modify the burrows of prairie dogs for their own year-round use. The burrows have entrances too small to admit coyotes or badgers, both of which pose serious threats to the swift fox. The coyote is their most serious threat, next to humans. Temporary burrows have single entrances and short tunnels and are usually located on high ground in shortgrass prairie, pastures, or even cultivated or plowed fields. A fan-shaped mound beside the entrance that consists of excavated subsoil serves as a lookout.

Up to five dens may be present in a 160-acre area, and the tendency of swift foxes to place their dens fairly near human habitations has contributed to their frequent destruction by poisoning. Natal dens have two or more entrances (up to at least nine have been reported), with one or two chambers that are usually 2 to 3 feet below the surface and about a foot in diameter. Evidently, females dig additional entrances to their burrows prior to whelping, and if other burrow entrances are closed during the whelping season, such as by plowing or disking, they are promptly reopened. Den entrances of swift foxes tend to average about 7.5 to 8.5 inches high and about 8 inches wide, with straight or curved tunnels

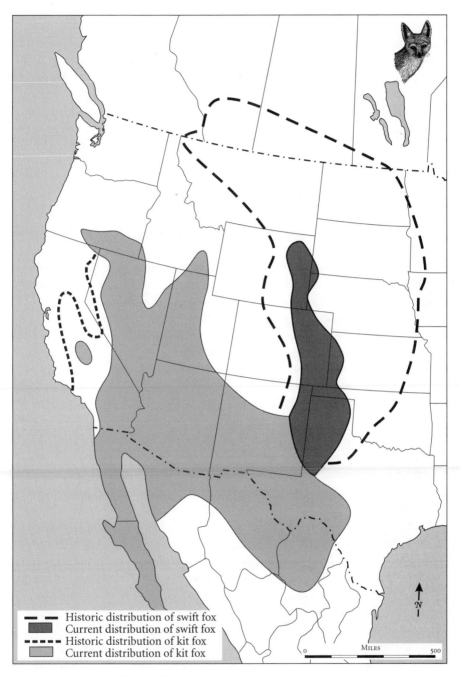

Distribution of swift fox and kit fox

leading to wider chambers. Sometimes badger burrows are also used as dens. In the closely related kit fox, a family group may use a cluster of up to seventeen dens, each having multiple entrances, with separate chambers for food caches and sleeping. By comparison, the black-tailed prairie dog usually has burrow entrances about 4 to 11 inches wide and are slightly narrower underground, so swift foxes may have difficulty proceeding very far into a prairie dog burrow without first enlarging it. Prairie dog burrows may also have as many as six entrances, but periodic plugging of some tunnels by the rodents may reduce the number of effective entrances at certain times, such as during the rearing of juveniles.

Although swift foxes often live near prairie dogs, they seem to consume them relatively rarely (Table 5) and cannot be regarded as a major threat. Rabbits and jackrabbits are swift foxes' primary mammalian prey, with a wide variety of rodents a close second. Lizards, ground-dwelling birds (horned larks, meadowlarks), and insects, especially grasshoppers, are also important items of diet, judging from Delbert Kilgore's research. In his Oklahoma panhandle study (1969), prairie dogs composed only 2.5 percent of the total prey occurrences in analyzed scat samples, with cottontails providing the most commonly encountered mammalian remains. Unidentified passerines were the most common avian remains, and lesser earless lizards were the most prevalent reptilian representatives. Among insects, *Melanoplus* grasshoppers composed about 30 percent of all the scat items found, and all the other insects found were also orthopteran insects. In a Kansas study by Marsha Sovada and others (2001), mammal remains appeared in 92 percent of scats examined, arthropods in 87 percent, birds in 24 percent, carrion in 23 percent, vegetation in 15 percent, and reptiles in 4 percent. In a Colorado study by David Anderson and others (1998), mammals composed 64 percent of estimated monthly mean food volume, arthropods 19 percent, and small birds 8 percent. It seems likely that insects are especially important during summer, as is true of burrowing owls, with mammals probably being taken more during colder periods, as the animals are opportunistic hunters.

Swift foxes have many enemies; however, the coyote is considered the largest single cause of their mortality, based on studies in Kansas, Wyoming, and elsewhere. Estimates of deaths caused by coyotes have ranged from 20 to 80 percent of all swift fox mortality. Coyotes reportedly do not eat the swift foxes they kill but simply kill them to reduce competition, just as wolves regularly kill coyotes. Swift foxes do not live long in the wild, probably averaging less than two years even for those that survive their vulnerable juvenile period. In a Wyoming study by Travis Olson and Frederick Lindzey (2002), annual adult swift fox survival rates ranged from 40 to 69 percent over a four-year period, with coyotes the primary cause of death. In a Colorado study (Kitchen, Gese, and Schauster

Table 5. Relative food item occurrences in scats of swift foxes from five states

FOOD ITEMS	SOUTH DAKOTA %[a]	NEBRASKA %[b]	OKLAHOMA %[c]	TEXAS %[d]	NEW MEXICO %[e]
Mammals					
Sciuridae	29.1	1.7	1.6	–	1.1
Leporidae	4.0	16.7	14.1	22.0	8.4
Cricetidae	9.4	19.4	10.1	5.6	0.1
Muridae	–	–	1.2	–	2.9
Heteromyidae	4.2	1.7	3.9	4.4	15.2
Geomyidae	1.4	1.1	–	–	0.3
Canidae	–	2.2	–	–	–
Soricidae	0.2	0.5	Trace	–	–
Mustelidae	0.7	–	Trace	–	–
Didelphidae	–	0.5	–	–	–
Carrion	0.1	11.1	–	–	0.6
Unidentified/Other	–	2.2	0.4	–	0.3
Birds	6.2	11.1	19.5	11.2	10.9
Reptiles	–	–	3.8	1.2	0.8
Fishes	–	–	–	1.2	–
Arthropods	26.9	16.1	41.3	36.8	53.2
Plants	12.9	15.6	–	17.6	4.0
Other	5.2	–	–	–	2.1

[a] Based on unspecified number of scats. Data from Uresk and Sharps 1986.
[b] Based on 52 scats (estimated 176 items). Recalculated from Hines and Case 1991.
[c] Based on 488 scats (estimated 975 items). Recalculated from Kilgore 1969.
[d] Based on 259 scats. Data from Cutter 1958.
[e] Based on 980 scats (estimated 1,482 items). Recalculated from Harrison 2003.

1999), 49 percent of twenty-five deaths of radiotracked swift foxes were caused by coyotes. In a Texas study by Jan Kamler and others (2003), in which survival rates ranged from 52 to 66 percent, the primary causes of deaths were vehicle collisions (42 percent) and coyote predation (33 percent). It is believed that golden eagles may also pose some threat to swift foxes, as these eagles have been known to kill both coyotes and domestic dogs, but little documentation of their danger to swift foxes exists.

With the extirpation of wolves from the Great Plains, which are not believed to have posed a serious threat to swift foxes, the coyote has become more widespread and abundant. Additionally, early cattle ranchers in the West who placed poison on buffalo carcasses to kill wolves often succeeded in killing the unsuspecting swift fox instead. It has been found that local control of coyotes tends to increase swift fox populations. Although the red fox is not large enough to pose a mortality threat to swift foxes, its aggressive behavior may drive the swift fox from areas occupied by red foxes. However, coyotes also drive out red foxes from an area, so a few coyotes in areas occupied by red foxes may actually benefit the swift fox.

The swift fox is not currently listed as either endangered or threatened nationally in the United States. Until January 2001, when its federal listing was terminated, the swift fox had been a candidate species for such listing (Category 2). In 1995 the U.S. Fish and Wildlife Service concluded that federal listing was warranted but was precluded by higher listing priorities for other species. The swift fox has a natural heritage listing of critically imperiled in South Dakota and Wyoming and is designated as state threatened in South Dakota. It is also state listed as endangered in Nebraska, where the largest numbers of county records since 1966 have occurred (in descending frequency) in Sioux, Dawes, Box Butte, and Kimball counties.

Trapping of swift foxes is still allowed in Kansas, with 181 reported legally taken between 1994 and 2002; the take averages about 22 per year. The Wyoming population is unknown but has been considered high enough to allow for some harvest, including live trapping and release in other states and Canada. The Colorado population of swift foxes was estimated at 7,000 to 10,000 animals in the late 1990s. Open seasons there between 1982 and 1991 resulted in annual harvests of 880 animals per year.

In Montana the swift fox was considered extirpated by 1978, but since 1998 more than 120 animals have been released in the state on Blackfeet tribal lands. It is considered a "species of conservation concern" in that state. Swift foxes have also been released since 2002 at the Turner Bad River Ranch in South Dakota, and in Badlands National Park, South Dakota. Several states have recently placed the swift fox under protection or listed it as a sensitive species. Studies over four years by Jan Kamler and others (2003) in northwestern Texas found that swift foxes there selected only shortgrass prairies and had lower than expected or complete avoidance of all other habitats, including Conservation Reserve Program grasslands. To ensure the swift fox's long-term survival, protection of native shortgrass prairies may be critical.

The swift fox is classified as nationally imperiled in Canada's national scheme for ranking endangered species. Swift foxes have been released since 1983 into Alberta and Saskatchewan in a major restocking effort, almost fifty years after the species was eliminated from Canada in 1938. Some of these Canadian animals have since spread into the boundary region of northern Montana.

The closely related kit fox is considered a sensitive species in region 3 (Southwestern Region) of the U.S. Fish and Wildlife Service, and the San Joaquin population of the kit fox is considered endangered under the Endangered Species

Fig. 22. Rodent associates of prairie dogs, including (top to bottom) hispid pocket mouse (adult), Ord's kangaroo rat (adult), deer mouse (adult dozing), and grasshopper mouse (adult).

Act of the United States. Little is known of the status of the Mexican race of the kit fox, but its current range is thought to be closely associated with the Mexican prairie dog's and may be limited to remnant grassland areas over a few hundred square miles of southwestern Nuevo Leon.

SMALL MAMMAL ASSOCIATES

Among the small mammal associates of the black-tailed prairie dog, the species most consistently present (Table 6) on prairie dog colonies are the thirteen-lined ground squirrel, the desert cottontail, black-tailed and white-tailed jackrabbits, the Ord's kangaroo rat, the deer mouse, and the northern grasshopper mouse (Fig. 22). The hispid pocket mouse, Wyoming pocket mouse, and western harvest mouse are less frequently associated species. All of these mammals are either rodents or lagomorphs, and except for the much larger jackrabbits, all are likely to be able to use a prairie dog burrow as a convenient escape route should aerial danger threaten. Only one of them is a hibernating species, the thirteen-lined ground squirrel, and abandoned prairie dog burrows undoubtedly make desirable hibernacula.

Using four criteria to try to establish various species' actual dependence on

Table 6. Vertebrate associates of black-tailed prairie dogs

REPORTED IN FIVE OF SIX STUDIES

Birds: Swainson's hawk, ferruginous hawk*, northern harrier*, prairie falcon, American kestrel*, mourning dove*, burrowing owl*, horned lark*, western meadowlark*

Mammals: thirteen-lined ground squirrel, badger, coyote

Reptiles: prairie rattlesnake

REPORTED IN FOUR OF SIX STUDIES

Birds: red-tailed hawk, golden eagle*, western kingbird*, barn swallow*, loggerhead shrike*, lark bunting*, chestnut-collared longspur*

Mammals: desert cottontail, deer mouse, pronghorn

REPORTED IN THREE OF SIX STUDIES

Birds: sharp-tailed grouse, mountain plover*, long-billed curlew*, upland sandpiper, common nighthawk, eastern kingbird, cliff swallow*, European starling, Red-winged blackbird*, brown-headed cowbird*, vesper sparrow*, lark sparrow*, McCown's longspur*

Mammals: black-tailed jackrabbit, white-tailed jackrabbit, Ord's kangaroo rat, northern grasshopper mouse, spotted ground squirrel

REPORTED IN TWO OF SIX STUDIES

Birds: mallard, northern pintail, green-winged teal, blue-winged teal, turkey vulture*, rough-legged hawk, greater sage-grouse, killdeer*, rock pigeon, great horned owl*, northern flicker, black-billed magpie, American crow, yellow-headed blackbird, Brewer's blackbird, grasshopper sparrow*

Mammals: plains pocket gopher, pocket gophers (*Thomomys* spp.), southern plains woodrat, swift fox, raccoon, long-tailed weasel, striped skunk, bobcat, mule deer, bison, domestic cattle, domestic sheep, domestic horse

Reptiles: ornate box turtle, lesser earless lizard, Texas horned lizard, sagebrush lizard, gopher snake, western diamondback rattlesnake

Amphibians: tiger salamander

REPORTED IN ONE OF SIX STUDIES

Birds: 53 additional species

Mammals: 15 additional species

Reptiles: 8 additional species

Amphibians: 9 additional species

Note: Listings are based on frequency of occurrence in six U.S. field studies (Reading et al. 1989).
*Bird species additionally seen at a Chihuahuan prairie dog complex (Manzano-Fisher et al. 1999).

prairie dogs for their well-being, Kotliar et al. (1999) found that nine vertebrate species fulfilled at least one of these criteria. The strongest evidence for such ecological dependence came from the black-footed ferret, the mountain plover, and the burrowing owl. The other six species (the ferruginous hawk, golden eagle, swift fox, horned lark, deer mouse, and grasshopper mouse) satisfied only one of the four criteria.

The badger, prairie rattlesnake, and tiger salamander failed to meet a single criterion. However, badgers are locally (e.g., in northern Mexico) attracted strongly to prairie dog colonies and seasonally may rely on them for a substantial percentage of their food. Small rodents, such as the northern grasshopper mouse and the deer mouse, are more abundant in prairie dog towns than in adjacent prairie, which may attract prairie rattlesnakes.

Temperatures within prairie dog burrows usually range from 5 to 10 degrees Celsius (41 to 50 degrees F) in winter and from 15 to 25 degrees Celsius (59 to 77 degrees F) in summer, allowing warmer than usual aboveground conditions in winter and cooler than general environmental temperatures in summer (Hoogland 1994). Warmer winter temperatures reduce stresses on both hibernating and active mammals during winter, and cooler summer temperatures reduce heat stress. Cooler temperatures also reduce water losses through evaporation from moist lung and skin surfaces, an important factor in the water-poor environments of the High Plains. Humidity within prairie dog burrows is also fairly high, averaging about 88 percent, which also helps greatly to reduce evaporation rates. More will be said of the mammal associates of prairie dogs in Chapter 6, including the large ungulates such as the pronghorn.

OTHER BIRD ASSOCIATES

All those birds [perhaps longspurs] are now setting, and laying their eggs on the plains; their little nests are to be seen in great abundance as we pass. There are meriads of small grasshoppers in the plains which no doubt furnish the principal aliment of this numerous progeny of the feathered creation. – *Meriwether Lewis, June 4, 1805*

All of the typical High Plains bird species are likely to be seen at one time or another in the vicinity of prairie dog towns (see Table 6), but the passerines most consistently seen there include the western meadowlark, horned lark, lark bunting, chestnut-collared longspur, and McCown's longspur (Fig. 23). Of these, the lark bunting and the McCown's longspur are essentially confined to shortgrass habitats, whereas the chestnut-collared longspur is usually found in slightly higher grass and mixed-grass vegetation. The horned lark and western meadowlark have notably broad ecological tolerances, ranging from desert grasslands to mixed-grass prairies, and so are not so closely associated with prairie dogs, but western meadowlarks do seem to preferentially occupy prairie dog colonies. All of these songbird species feed small grasshoppers and other ground-dwelling insects to their own young, so the low-clipped and newly growing grasses and forbs associated with a prairie dog colony probably make

Fig. 23. Avian associates of prairie dogs, including McCown's (male, upper left) and chestnut-collared (male, upper right) longspurs, lark bunting (male, middle left), horned lark (adult, middle right), and mountain plover (adult, bottom).

for rich foraging opportunities. Some insectivorous birds even briefly enter prairie dog burrows, apparently to feed on arthropods.

Among shorebirds the mountain plover is the species most consistently found around prairie dog towns, largely because it prefers very short grasses and bare ground for foraging and nesting. The upland sandpiper, killdeer, and long-billed curlew are also sometimes found near prairie dog towns, with the killdeer usually being associated with bare-ground, pebble-rich habitats, and the curlew and upland sandpiper much more characteristic of mixed-grass prairies. Aerial-feeding insectivore species, such as eastern and western kingbirds, cliff swallows and barn swallows, and common nighthawks also find the areas above prairie dog towns productive for finding insect foods. The killdeer, horned lark, mourning dove, and barn swallow have all been reported as more common in prairie dog colonies than in surrounding mixed-grass prairies. More will be said of these passerine and shorebird associates of prairie dogs in Chapter 6.

Small raptors, such as the loggerhead shrike and the American kestrel, are attracted to the diverse foraging opportunities in prairie dog towns. These species also need elevated sites for perching while scanning, and for nest sites they need either small trees (shrikes) or good-sized trees with cavities (kestrels). As a result neither species is likely to nest within the limits of most prairie dog colonies, but they often include such colonies within their foraging ranges. The larger raptors (hawks, eagles, falcons, and owls) associated with prairie dog towns are numerous and will be discussed in Chapter 7.

HERPTILE ASSOCIATES

In this neighborhood are many villages of the prairie dogs (Arctomys ludovicianus Ord) in the abandoned burrows of which rattlesnakes abound. It has been affirmed that these two species live peacefully together in these burrows; but observers of nature have proved that the snakes take possession of abandoned burrows only, which is the usual course of things. – *Prince Maximilian of Wied*, Travels in the Interior of North America, 1832–1834 *(1906)*

One of the many attractive if patently false beliefs about the American West is that prairie dogs somehow live in peace with rattlesnakes in their burrows, an idea that is nearly two centuries old but that has no basis in fact. Rattlesnakes have few if any friends on the prairie – and certainly not prairie dogs – but they have plenty of enemies. Surprisingly few natural encounters between rattlesnakes and prairie dogs have been documented, although rattlesnakes are largely nocturnal and so such encounters are likely to go unseen by humans. However, "staged" encounters involving various species of snakes have been experimentally arranged.

Zuleyma Halpin (1983) described four naturally occurring encounters that he observed on Quivera National Wildlife Refuge in Kansas, where rattlesnakes, bullsnakes, prairie kingsnakes, and eastern yellow-bellied racers all occur and represent potential prairie dog predators. Of the four encounters observed, one involved a prairie kingsnake, and at least two and probably three involved bullsnakes. The most common response to seeing a snake was repeated jump-yipping, in one measured instance at the rate of about one display every five seconds. In the case of an adult female with a juvenile nearby, the female performed rapid head bobbing and sharp vocalizations that resembled the first syllable of the jump-yip display. The snake was situated in a single-entrance burrow, and the female repeatedly kicked dirt into the burrow with her hind legs, occasionally turning and pushing more in with her front legs and tamping it down with her nose. After twelve minutes the burrow had been completely covered over and the female walked away from it. A large, unhurt bullsnake was later dug out of the sealed entrance by Halpin.

Using both tethered rattlesnakes and bullsnakes, W. J. Loughry (1987) has studied the responses of black-tailed prairie dogs to these predators. He found

Fig. 24. Herptile associates of prairie dogs, including Texas horned lizard (top), tiger salamander (middle), and ornate box turtle (bottom).

that their behavior varies with the type of snake involved, with the age, sex, and parental status of the prairie dogs, with the time of year, and with whether the encounter was experimentally staged or naturally occurring. He found that adult prairie dogs were more likely to harass snakes than were juveniles, regardless of regional location. Some regional differences in adult behavior between prairie dogs in Texas and South Dakota were observed, with mothers in South Dakota calling more and staying closer to snakes than mothers in Texas, and with no sexual differences observed between adult male and female behaviors toward snakes in South Dakota or between the responses of fathers and nonfathers, contrary to what had been observed in Texas. Loughry concluded that the history of a local population as to its prior exposure to snakes may result in substantial differences in their present-day responses to them.

Several other reptiles, such as the Texas horned lizard, tiger salamander, and ornate box turtle (Fig. 24), have often been documented in prairie dog colonies, but none of these species poses a threat to prairie dogs. The horned lizard probably is attracted to the ants that are abundant around prairie dogs, and the tiger salamander and box turtle are likely to use the burrow system of the prairie dogs as fairly safe and comfortable retreats from environmental extremes.

6

Other High Plains Wildlife

Born on the American Steppes

We found the Antelope extremely shye and watchfull insomuch that we had been unable to get a shot at them; when at rest they generally seelect the most elevated point in the neighborhood, and as they are watchfull and extremely quick of sight and their sense of smelling very acute it is impossible to approach them within gunshot. . . . I think I can safely venture the assertion that the speed of this anamal is equal if not superior to that of the finest blooded courser. – *Meriwether Lewis, September 17, 1804*

PRONGHORN

Few North American large mammals are more beautiful than the pronghorn. Its dark eyes are as large and lustrous, and its eyelashes as long and conspicuous, as those of any 1950s silver screen goddess (Fig. 25). Its horns are an evolutionary compromise between the permanent, constantly growing true horns of cattle-like animals and the bony antlers carried by males of the deer family, which are seasonally shed and entirely regrown each year. Adult pronghorns of both sexes carry horns, which have flattened bony cores covered by a black fibrous sheath that is shed annually after the breeding season. The horns of females are smaller than those of adult males and lack the males' distinctive forks, or "prongs." The pointed horns of males are certainly used in male-to-male dominant battles and can cause serious injuries among males when they are competing for females. The function of the females' smaller horns is less obvious. They may serve as defensive weapons against possible predators, but female pronghorns defending their newborn young from coyotes are more likely to kick the coyotes than to try impale them with their small horns. And given the opportunity, pronghorns have long relied on their superior speed to avoid and escape all enemies, rather than facing them and using direct defensive maneuvers.

If the golden eagle is the symbolic king of the High Plains skies, then the pronghorn deserves at least the title of prince of the High Plains themselves. Yet our beautiful golden eagle is as much at home in Europe (where it is often the

Fig. 25. Adult male pronghorn.

icon of royalty) as in North America, and even the American bison can trace its ancestry back at least three hundred thousand years to the ancestral Siberian steppe bison, *B. latifrons*. However, the pronghorn is a purely North American evolutionary product, with a historic range that once encompassed much of the American West. That range now has been reduced to many small, isolated population units, and its eastern limits have retreated steadily westward (Map 8). It has also suffered at the southern end of its range, with a race limited to Mexico's Baja Peninsula and a Sonoran Desert race from northwestern Mexico and extreme southern Arizona, each listed as endangered by the World Conservation Union. The pronghorn's remaining range is still generally centered around the range of various large sagebrush species, especially big sagebrush. Big sagebrush is a drought-tolerant evergreen shrub that cattle ranchers regard as having only fair to poor forage because of its astringent oil content, but which is critical to pronghorns as a primary source of winter food. Not only are its winter leaves high in protein, but the woody shrub also is tall enough to protrude above the snowline in most winters.

Aided by their enormous eyes, pronghorns can detect a moving human or coyote from a mile or more away, but they also tend to ignore all stationary objects in their environment. Sometimes to their peril, they will approach to

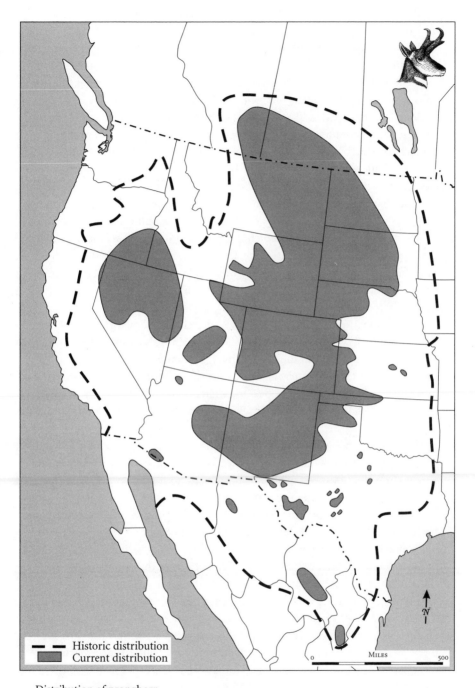

Distribution of pronghorn

Historic distribution
Current distribution

investigate an unfamiliar object seen in the distance, such as a crouching hunter. The pronghorn's ears are unusually long and pointed, and its hearing is remarkably acute. Its large ears, like those of the kit fox, suggest it may have evolved in a relatively warm climate. Likewise, the pronghorn has a very long muzzle and an associated keen sense of smell, as Meriwether Lewis astutely observed. Its double hooves are deerlike, with the two halves spreading apart anteriorly to increase the amount of ground area contacted while the animal is running, and the bottom of each hoof has cartilaginous padding for cushioning foot impacts.

Like members of the distantly related deer family, pronghorns have no clavicles (collarbones), which allows the shoulder blade to shift forward unimpeded and increases the effective stride of the forelegs by as much as 20 percent. Uniquely among American ungulates, pronghorns resemble cheetahs and cats by having a flexible backbone, which during fast running allows the muscular hind legs to reach forward until they are momentarily ahead of the forelegs. Then the spinal muscles contract, straightening the back, propelling the animal ahead, and allowing the forelegs to gain a maximum distance forward before they again touch the ground and begin a second stride.

An additional speed-gaining adaptation present in pronghorns and deer is the "springing ligament," an elastic tendon in each leg that extends from the bottom of the toes around the back of the foot joint to the rear surface of the shinbone. When weight is placed on the toe, the tendon is stretched forward into a position that allows the hooves to be flat on the ground, but when weight is released, the hooves snap down and back, providing an extra jolt of forward and upward energy.

The pronghorn is generally accepted to be the fastest runner of all North American mammals, regularly cruising at about 40 miles per hour and attaining a maximum clocked speed of about 54 miles per hour when running on level, hard ground. At such times the animals may make horizontal leaps of 10 to 20 feet, but they do not attempt vertical jumps; a pronghorn would much rather try to run under a shoulder-high wire fence at full speed than try to jump over it. Their aversion to such jumping is so great that a pronghorn will make a mile or more detour in order to go around the end of a fence that it cannot get under rather than try simply to jump over it.

Very large lungs and a correspondingly large heart provide the oxygen needed for the animal's prodigious running and endurance abilities. Runs of up to about 3 miles are common, and over the course of a year, pronghorns may move as far as 100 miles between their summer and winter ranges. Although pronghorns drink water where available, they can also obtain the moisture they need from the plants they consume. Summer and autumn home ranges in Wyoming average about 1 to 2 square miles.

It is well known that pronghorns are attracted to prairie dog towns. Like bison they are fond of the newly growing and nutritious green vegetation characteristic of such locations. In studies of the foraging ecology of pronghorns and bison in black-tailed prairie dog towns, Kirsten Krueger (1984, 1986) reported that, in sequence from a prairie dog town center to its edges and beyond, the percentage of grasslike plants increases, as do estimates of mean total vegetation cover and mean aboveground vegetation biomass, whereas the incidence of forbs and shrubs decreases. Nitrogen concentrations of western wheatgrass were observed to decrease in the same progressive center-edge-offtown sequence, and studies by D. Coppock and others (1980) have similarly found that shrub and forb (broad-leaved herb) nitrogen levels are highest in town centers. Bison tended to feed heavily (85 to 99 percent of all observed use) on the town edges that are dominated by grasslike plants, whereas 57 to 97 percent of all the pronghorn feeding observed in dog towns occurred in town centers. And, whereas bison fed to a larger degree than expected by chance among on grasslike plants (relative to their quantitative availability), pronghorns selected forbs at rates three to ten times their relative availability.

Prairie dogs feeding near town centers foraged much like pronghorns, but those feeding on town edges had foraging patterns similar to those of bison, suggesting that they were less selective in foods taken than were either pronghorns or bison. Pronghorns as well as prairie dogs were prone to make food discriminations that favored the more digestible plant components, selecting for consumption leaves, and sometimes also flowers, over plant stems. Bison and pronghorns both made heavy use of prairie dog towns, but their significantly different usage patterns in terms of foraging areas preferred and plant types selectively utilized would tend to reduce their interspecies competition. Bison and pronghorns showed the highest usage rate of prairie dog towns during summer months of active vegetational growth, but their overall observed use of these towns was always much higher than would have been expected by chance alone. The prairie dogs seemingly did not benefit directly from the presence of pronghorns on their towns, and there may have been very slight negative behavioral (avoidance) effect from their presence, based on nearest-neighbor data.

Besides providing a source of unusually nutritious foods, the bare soils of the dog towns may offer wallowing and dusting opportunities for large ungulates, such as bison and pronghorns; the open spaces associated with these towns provide good visual observation points for locating predators; and the effective warning system of the prairie dogs may also benefit the larger animals. Prairie dog towns are certainly not predator free; instead, Krueger (1984, 1986) found that they had an incidence of coyotes and badgers 5.7 times greater than that typical of off-town grassland sites.

Fig. 26. Shortgrass ground squirrels, including (top to bottom) spotted (adult), thirteen-lined (hibernating adult), and Richardson's (adult in upright-alert posture).

RICHARDSON'S AND THIRTEEN-LINED GROUND SQUIRRELS

Carrying forward our analogy of the pronghorn as the royalty of the American High Plains, the thirteen-lined, Richardson's, and spotted ground squirrels (Fig. 26) would then seem to fit the category of High Plains commoners, along with the black-tailed prairie dog. Like human commoners in hierarchically organized societies, they are highly abundant, industrious, and relatively short-lived. The Richardson's ground squirrel's range does not extend as far south as that of the black-tailed prairie dog, and the spotted ground squirrel is a shortgrass and desert grassland species. The range of the thirteen-lined ground squirrel extends from the shortgrass to tallgrass prairies and even south to the coastal grasslands of eastern Texas.

Together with several other ground squirrels, these species complement the black-tailed prairie dog as basic herbivore components of the grassland ecosystems of interior North America. As such, ground squirrels are near the bottom of the grassland food chain and, together with rabbits and hares, make up much of the biomass base on which the higher-level carnivores, such as coyotes and foxes, must depend for maintaining their own populations.

The thirteen-lined ground squirrel was found to be present on prairie dog towns in five of six separate studies; the spotted ground squirrel, in three; and the Richardson's ground squirrel, in one (see Table 6 in Chapter 5). Ground squirrels of all these species favor the shortgrass environment typical of prairie dog towns, perhaps at least partly because of the excellent visibility provided. Ground squirrels may also be attracted to prairie dog towns because of the availability of nutritious vegetation, the potential for exploiting the effective prairie dog predator alarm system, and the abundance of escape holes should a predator suddenly appear. The same reasons for attraction to prairie dog towns might also apply to other rodent associates, such as the deer mouse, Ord's kangaroo rat, and northern grasshopper mouse (see Fig. 22 in Chapter 5), which all were reported on prairie dog towns in at least three of six studies. Of these the northern grasshopper mouse is something of an outsider, since it is mostly carnivorous and may be attracted to grasshoppers and other insects typical of the dog towns rather than to the vegetation growing there.

MOUNTAIN PLOVER

The mountain plover is a medium-sized, somewhat nondescript shorebird lacking any contrasting colors and plumage patterns; it somewhat resembles a killdeer that has been out in the sun too long. Unlike the broadly tolerant and widespread killdeer, mountain plovers have a particular attraction to very short grass (up to only about 2 inches high) and bare ground. They are largely confined to heavily grazed grasslands, although in Montana a study by S. Olson-Edge and W. Edge (1987) found that cattle grazing alone, in the absence of prairie dog downs, did not attract mountain plovers. A combination of short grasses and low litter accumulation does attract them, such as in areas where confined bison are pastured, around stock tanks, and where sheep are being herded. However, areas of long-term overgrazing are avoided. In many ways the mountain plover and the swift fox have been caught up in the same net of human exploitation and reduced shortgrass plains habitat, which has nearly spelled extinction for both of them. Like the swift fox, the mountain plover's range is a fraction of its former distribution (Map 9).

In several areas mountain plover nesting has been found to be positively related to prairie dogs. A remnant nesting colony in the Uinta Basin, northeastern Utah, was found by Ann Ellison and Clayton White (2001) to be directly associated with white-tailed prairie dogs, and in Colorado as well as Montana a clear positive association exists between nesting mountain plovers and the presence of black-tailed prairie dogs. In Montana, prairie dog towns smaller than about 25 acres were found to provide only marginal plover breeding habi-

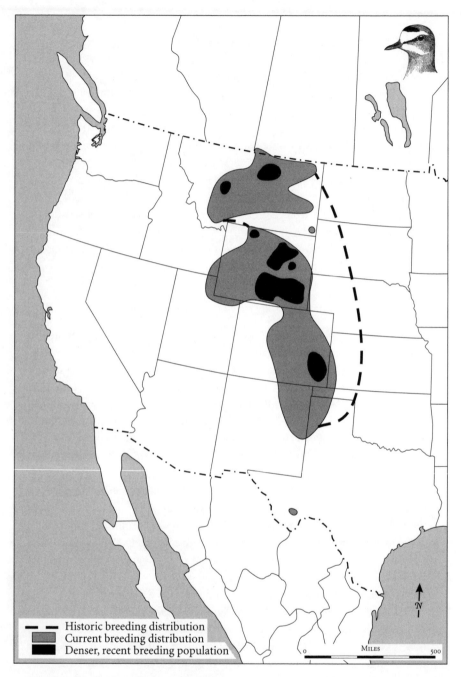

Breeding distribution of mountain plover

Fig. 27: Heads and bills (top to bottom) of long-billed curlew (female above), marbled godwit (female above), upland sandpiper (sexes alike), and mountain plover (sexes alike).

tat, and the birds avoided breeding in areas without prairie dog towns that were studied by Craig Knowles and others (Knowles, Stoner, and Gieb 1982). The highest densities of breeding plovers occurred on dog towns of large size. The mean size of towns used by them was 142 acres, according to a study by Craig Knowles and P. Knowles (1984). Prairie dog towns apparently provide greater food resources and more vulnerable prey (insects) than do areas outside dog towns, according to studies by Sally Olson (1985). The short, stout beak of the mountain plover adapts it well to picking up and crushing insects such as grasshoppers, the beak being both shorter and more robust than that of the upland sandpiper, which forages in a similar way but in taller grassy habitats. Both in turn have shorter beaks than curlews and godwits, which often forage by probing in shallow water (Fig. 27).

Like many grassland birds mountain plovers use aerial displays to advertise their breeding territories. One of the spectacular aerial displays is the "falling leaf," in which a male rises to a height of 15 to 35 feet and, at the apex of the flight, sets his wings at a sharp upward angle and then floats back to the ground, rocking back and forth and calling loudly (Fig. 28). A much less commonly

Fig. 28. Male aerial displays of four prairie birds, including (top to bottom) horned lark, upland sandpiper, marbled godwit, and mountain plover. Vocalizations are based on published sonagrams.

seen aerial display is the "butterfly flight," during which the male flies over his territory with deep and slow wingbeats, somewhat resembling a giant butterfly or moth.

Breeding territories of mountain plovers average about 40 acres in Colorado, with some spatial overlapping at the edges during the brood-rearing phase. It has also been judged by Fritz Knopf and J. Rupert (1996) that mountain plovers need at least 70 acres to raise a brood. Areas that usually support mixed-grass prairie can be made attractive for mountain plovers by performing heavy grazing in summer or late winter, but extreme overgrazing should be avoided.

The mountain plover is currently listed as an endangered species in Canada and as threatened in Mexico. In 1993 the mountain plover was listed as a Category 1 Candidate Species under the U.S. Endangered Species Act. Its federal status has recently (2003) been downgraded, as the U.S. Fish and Wildlife Service has concluded that it is not warranted for listing as a nationally threatened or endangered species, although its numbers continue to decline. This change in status is under legal appeal. The mountain plover has been ranked by the Natural Heritage Program as extinct in South Dakota, as critically imperiled in Nebraska, and as imperiled in Colorado.

LONG-BILLED CURLEW, MARBLED GODWIT, AND
UPLAND SANDPIPER

Three species of North American shorebirds bear the authentic hallmarks of the North American mixed-grass prairies. In sequence of diminishing bill lengths and overall body sizes (the usual coarse-grain measure of all shorebirds), they are the long-billed curlew, the marbled godwit, and the upland sandpiper. Each is a bird that any shorebird lover might well personally select as the most beautiful bird in the world, for each offers an unforgettable combination of outward appearance and fascinating behavior patterns that makes every encounter an event to be remembered for life.

Of the three, the long-billed curlew (see Fig. 27) is perhaps most closely tied to the High Plains grasslands, but it also extends west into the Great Basin arid grasslands, the bunchgrass prairies of the Columbia Basin, and the rain-shadow region of the Cascades and Sierra Nevadas. Its centers of abundance in the Great Plains are in the cool prairies lying east of the Rockies in Alberta and Montana, in the Nebraska Sandhills, and in the dry prairies situated between the Arkansas River of eastern Colorado and western Kansas and the Canadian River of eastern New Mexico and northwestern Texas.

Besides being the largest of the prairie shorebirds, the long-billed curlew exhibits the biggest sexual differences in the external appearance of adults, especially as to their differing bill lengths and bill curvatures. Females are larger (by about 15 percent in weight) than males and exhibit bills up to 9 inches long, which are noticeably decurved in almost a new-moon shape, whereas the bills of males are about 6 inches long and show less curvature. These gender-based bill differences probably produce some ecological segregation in foraging. All curlews have variously lengthened and decurved bills, which presumably are useful when probing wet sand along shorelines, as often occurs during winter, but during summer little such probing is apparent, even in the sand-rich Nebraska Sandhills region.

Long-billed curlews seek out shortgrass areas for nesting, often on gently rolling terrain and where the birds are able to hide by crouching but still are able to see out from above the vegetation with the neck erect. In densely used areas, nesting territories reportedly average about 35 acres, but various estimates have ranged from as little as 15 to as much as 50 acres. Topography may influence a male's territory size, with diverse topography and varied habitat supporting smaller territories and denser populations than do open, flatter, and less diverse habitats.

Long-billed curlews were reported in four of seven studies reporting vertebrate associations in prairie dog towns. In a Chihuahuan prairie dog town they were the most common shorebird reported and made up 16 percent of all 6,270 birds seen during more than a year of study. The long-billed curlew is listed as vulnerable in Canada and is considered "highly imperiled" in the U.S. Shorebird Conservation Plan.

The marbled godwit (see Fig. 27) is effectively limited as a breeding bird to the glaciated mixed-grass prairies of the northern plains, from Alberta south and east to South Dakota. It is most abundant from southern Alberta to southwestern Saskatchewan and in the prairie pothole region of central North Dakota, where it often breeds in the company of other mixed-grass and wet-meadow shorebirds, such as upland sandpipers, willets, and Wilson's phalaropes. Marbled godwits prefer to breed in areas having both short and moderately dense upland vegetation as well as nearby wetlands, especially temporary alkali ponds. Large areas of contiguous grasslands, of at least 250 acres, are preferred, and estimated breeding territory sizes alone average nearly that large.

As with the long-billed curlew, adult female marbled godwits are about 20 percent larger than males and have longer, if slightly less colorful, bills. Both marbled godwits and long-billed curlews have beautiful, cinnamon colored underwing coverts, which are visible during aerial display and also when the birds raise their wings in ground display, seen especially just after landing.

Marbled godwits have sometimes been seen in prairie dog towns; they were reported in one of seven studies of prairie dog associated vertebrates. They do not appear to be as highly attracted to prairie dog towns as are upland sandpipers and long-billed curlews.

The upland sandpiper (see Fig. 27) is a Neotropical migrant, summering and breeding in grasslands from western Alaska to the Atlantic coastal plain but reaching its greatest abundance in the mixed-grass prairies from central North Dakota south to east-central Kansas. It is the smallest of the three species discussed here, and the females are only about 5 percent larger than the males. In contrast to the large curlews, the adult sexes cannot be distinguished from their outward appearance alone. Yet the upland sandpiper in many ways seems to represent the end point of a morphological and evolutionary gradient that begins with the largest curlews, extends through the smallest curlew species, and ends with the upland sandpiper.

The upland sandpiper requires extensive areas of shortgrasses for breeding, probably where the grasses are short enough for the bird to see over but long enough for it to crouch and hide. Grasses 6 to 10 inches high seem to be the preferred vegetation height for foraging birds, with nesting birds seemingly preferring vegetation at least 8 inches high. Nests are often placed in slight depressions, usually where grasses arch overhead to help conceal the bird from above. Grass clumps, dense vegetation, or the bases of forbs may also serve as nest sites. Areas moderately grazed by cattle, or grazed by bison and also burned, have been reported as preferred by upland sandpipers for both breeding and foraging. Grasslands with areas of short vegetation (6 to 12 inches) for foraging, slightly taller vegetation for nesting, and short to medium vegetation (over 8 inches high for older broods) for brood rearing are probably ideal, especially if the grassland areas are at least 250 acres in size. Individual territories have been estimated as 20 to 30 acres, but grasslands of fewer than 250 acres support few if any upland sandpipers.

Like mountain plovers, the long-billed curlews, marbled godwits, and upland sandpipers all perform aerial territorial displays high above ground, whistling loudly. The vocalization of the upland sandpiper, often uttered while flying high above its territory, is notably loud and melodious. It sounds much like the "wolf whistle" sometimes used by men when trying to attract a woman's attention (see Fig. 28). When upland sandpipers land, often on a fence post, they briefly hold their wings vertically above the back, exposing the beautiful barred, black-and-white underwing pattern.

Upland sandpipers were seen in prairie dog towns in three of seven studies that documented vertebrate prairie dog associates. The upland sandpiper is listed as endangered in four eastern states and as threatened in four others. In the Great Plains region it is considered a species of special concern in Wyoming, is listed as rare in Colorado, and is on the "sensitive species" list for Region 3 of the U.S. Forest Service.

HORNED LARK AND LARK BUNTING

Horned larks are widespread across the grasslands and tundras of North America, with various subspecies being adapted to differing climates and vegetational conditions. The western High Plains support large numbers of horned larks. This species was the most numerous of all the grassland birds seen by Patricia Manzano-Fischer and others (Manzano-Fischer, List, and Ceballos 1999) at eight localities within a large prairie dog complex in northern Chihuahua over a period of more than a year. Of 6,270 birds seen, the horned lark composed 65 percent of the total. Lark buntings were the second most common, making up 15 percent of the total.

Horned larks are insectivorous throughout the warmer months, taking animals and animal materials almost exclusively during summer, especially grasshoppers, beetles, and the larvae of butterflies and moths. During fall the incidence of vegetational materials rises, with seeds making up 60 to 70 percent of the diet, and this percentage increases to nearly 100 percent in winter. Grass seeds make up the majority of plant foods, with the seeds of forbs important locally. Such an environment makes prairie dog towns favorable for foraging.

Territory sizes are generally about 4 acres in the Midwest, with somewhat smaller ones reported from the shortgrass prairies of eastern Colorado. In Colorado the horned lark is believed to be the state's most abundant breeding bird, with 2.7 million pairs, concentrated on the eastern plains. In one Weld County study two-thirds of the nests were in shortgrass prairie, 17 percent were on bare ground, and the rest were in various other habitat types (Kingery 1998).

Like nearly all grassland birds, male horned larks perform song-flights high above their territories, sending forth a stream of melodious notes as they fly up to several hundred feet above ground (see Fig. 28). The song is uttered as the wings are set and the tail spread, with the bird slowly losing altitude. At the end of the song the male regains altitude and repeats the performance. Typically it faces into the wind while singing, and may remain nearly stationary if the breeze is adequate. Ending its song the bird closes its wings and plummets back to earth silently. Such song-flights may last up to eight minutes but are usually much shorter.

The lark bunting is similarly characteristic of the shortgrass plains. In Colorado it has been judged the fourth most common breeding bird, with an estimated 1.6 million pairs, concentrated in the eastern plains, although during recent drought years the lark bunting has been losing ground rapidly. It is often found breeding in the same locations as the horned lark and the two longspurs, but the larks and longspurs (especially the McCown's) occupy more barren habitats, and the horned lark starts nesting a month sooner than the lark bunting and continues longer, feeding its young mostly beetles. The two longspurs also start sooner and end earlier. Lark buntings feed larger prey (mostly grasshoppers) to their young, while longspurs feed them a mixture of small grasshoppers, beetles, and the larvae of butterflies and moths (Kingery 1998). Probably both longspurs and the lark bunting evolved in the presence of bison, other ungulates, and prairie dogs, favoring grazed pastures over sites with taller vegetation.

Like the horned lark and the two longspurs, the lark bunting has a wonderful song-flight display (Fig. 29), during which it ascends into the sky, sets its wings into a fairly sharp V-shape, and floats downward while rocking from side to side, singing loudly. Once this display has been seen and heard, the shortgrass plains will never again seem the least bit boring to a naturalist.

Fig. 29. Male aerial displays of the McCown's longspur (top), chestnut-collared longspur (lower left), and lark bunting (lower right). Vocalizations are based on published sonagrams.

MCCOWN'S AND CHESTNUT-COLLARED LONGSPURS

Four species of longspurs exist in North America, all associated with either grasslands or tundra, but only the McCown's and the chestnut-collared breed on grasslands south of Canada in the High Plains region. All four species are seed eaters, with typical heavy bills adapted for crushing seeds, and all feed insects to their young during the brood-rearing season. Even during the breeding season adults eat a high incidence of grass seeds, the amounts being estimated by various studies at 64 to 74 percent for the McCown's and 28 percent for the chestnut-collared. Outside the breeding season the diet becomes almost exclusively based on seed eating.

All the longspurs have a considerable amount of white in the tail feathers, and all of them perform spectacular flight displays over their breeding territories (see Fig. 29). Typically the male flies quickly upward, sometimes circling over his territory. On reaching an apex, the male spreads his tail and sings loudly while descending with the wings widespread, exposing their whitish undersurfaces. During this final singing phase the bird may either hold its wings rigidly outward

(McCown's) or perform some wing flapping during the descent (chestnut-collared).

Longspurs of uncertain species were the third most numerous of all the passerine birds seen by Patricia Manzano-Fischer and others at the above-mentioned prairie dog complex in northern Chihuahua (Manzano-Fischer, List, and Ceballos 1999). Of 6,270 total birds seen, longspurs composed 3.5 percent. These longspurs were almost certainly McCown's or chestnut-collared (or both), whose winter ranges center in Texas and the northern Mexican Plateau, the McCown's favoring somewhat more arid sites than does the chestnut-collared.

I observed also a small bird [McCown's longspur] which in action resembles the lark. . . . This bird or that which I take to be the male rises into the air about 60 feet and supporting itself in the air with a brisk motion of the wings sings very sweetly, has several shrill short notes reather of the plaintive order which it frequently repeats and varies, after remaining stationary about a minute in his aireal station he descends obliquely occasionally pausing and accomnying his descension with a note something like twit twit; on the ground he is silent. – *Meriwether Lewis, June 4, 1805*

"We don't need no ecosystem around here." (Response of one local Wyoming individual to applying the concept of ecosystem management to the Greater Yellowstone region) – *Quoted in Seymour Fishbein,* Yellowstone Country: The Enduring Wonder *(1989)*

7

The High Plains Raptors

Death Rides on Silent Wings

It may be said that [in the minds of the Oglala Dakotas] the eagle holds priority over all non-human living beings, because the eagle "flies higher than other birds," "sees everything," and "moves through the skies in the sacred form of the circle." – *Joseph E. Brown*, Animals of the Soul *(1992)*

The avian High Plains predators of prairie dogs are many, the most important species including golden and bald eagles, ferruginous hawks, and prairie falcons. At least occasionally, red-tailed hawks, northern harriers, and peregrine falcons may also take prairie dogs, especially younger animals. Prairie dogs may also at times be killed by several other larger raptors, including the occasional daytime-hunting great horned owl and the vagrant goshawk. The lives of prairie dogs are not safe either by day or by night, for they may be under attack both from land and by air.

GOLDEN EAGLE

For many people golden eagles (Fig. 30) perfectly exemplify the beauty and grandeur of the western plains. Their black-tipped, white tail feathers decorated the war bonnets of such Native American tribes of the High Plains as the Dakotas and the Cheyennes, and whistles made from their hollow wing bones were used by warriors to gather courage when riding into battle. For modern-day naturalists a golden eagle soaring high overhead on invisible thermals can lift the spirit of nature lovers as few other birds can. Golden eagles are often found near prairie dog towns in both summer and winter, where free meals are served up regularly in the form of young or unwary animals that stray too far from the safety of their burrows.

Golden eagles are rabbit-hare (lagomorph) and ground squirrel–prairie dog (sciurid) predators par excellence throughout their entire North American range. In a survey of North American food studies by Mike Kochert and others (2002), representing over twelve thousand identified prey items from fourteen geographic regions, rabbits and hares constituted 3 to 89 percent of all identified

Fig. 30. Adult golden eagle.

prey and in eight of the fourteen regions accounted for more than half the total prey items. Sciurid rodents – especially ground squirrels, marmots, and prairie dogs – represented from 2 to 68 percent and in six of the fourteen regions made up more than 30 percent of the total prey items. Together these two prey groups represented from 49 to 94 percent of all prey items identified and represented at least 70 percent of the total in ten of the fourteen regions. Prairie dogs were locally most important in southeastern Montana and northern Wyoming, where they composed 58 percent of 312 identified prey items; in southeastern Wyoming, where they represented21 percent of 721 items; and in New Mexico and western Texas, where ground squirrels (rock squirrels) and prairie dogs collectively made up 16 percent of 1,396 items.

The eastern edge of the golden eagle's breeding range in the Great Plains corresponds closely to the eastern edge of the black-tailed prairie dog's and jackrabbit's ranges. During winter the golden eagle's range limits contract somewhat westward. Throughout this period hibernating ground squirrels become seasonally unavailable, probably making prairie dogs, rabbits, and hares a relatively more important part of the species' prey base. Because ground squirrels and prairie dogs typically contain substantially more body fat than jackrabbits, they also provide more energy per unit weight of the prey. The average prey weight of the golden eagle is 1.3 pounds, but animals as large as nearly 13 pounds have been reported, including young and some adult ungulates. Domestic animals that have been reported as prey include calves, sheep, goat, pigs, dogs, and cats, but many of the reports of eagles eating sheep and cattle refer to carrion consumption.

Golden eagles are faithful to their breeding territories as well as their general wintering areas. Both are chosen probably in large part on the basis of reliable food supplies as well as for the presence of suitable perching sites and nesting locations. In some areas wintering densities of golden eagles have closely correlated with the abundance of jackrabbits, and similarly long term declines in breeding densities have been correlated in the Snake River Canyon region of Idaho with corresponding reductions in jackrabbit populations and amounts of shrub cover. Also, the percentage of territorial pairs laying eggs, their mean nesting success rate, and the average size of their broods at fledging have all been positively correlated with local jackrabbit abundance in Idaho. Likewise, nonbiological factors such as severe winter weather that causes excessive energy drain, or thermal stress on chicks produced by hot weather during the first six weeks after hatching, can also adversely affect golden eagle breeding success. The mean distance between adjacent occupied nests of golden eagles in Wyoming has been found to average about 3.3 miles, based on a sample of twelve nests. However, a single pair may maintain and repair as many as fourteen nests within their territory as part of their courtship behavior. Breeding home ranges in Wyoming have been estimated to range from 10 to 21 square miles.

During fall some golden eagles in the Great Plains drift southward, with immatures moving south earlier than adults. However, many pairs on the northern plains remain within their breeding territories, especially when prey remains available. The densest winter populations occur in the Wyoming Basin, centering around Casper, Wyoming, and extending west to the Green River Valley of Wyoming. Winter foods are largely jackrabbits and cottontails, as they also are in summer, but – at least locally – ungulate carrion may play a somewhat larger role in the winter diet than it does during summer.

When hunting, golden eagles use a variety of attack methods. They typically approach from upwind, often after soaring in a thermal, then attack by stooping from a high angle. Low-angle attacks are more likely when ground-dwelling mammals are the target (Fig. 31); a high-angle stoop may be used when flocking or slow-flying avian prey, such as geese or cranes, are involved. Golden eagles also hunt prairie dogs and ground squirrels using "contour flights," during which they approach their prey in low-altitude flights just above ground level, using land conformation variations to hide their approach, in a strategy more often used by prairie falcons.

At least in nonmigratory populations, golden eagles maintain their pair bonds year-round and sometimes hunt cooperatively. For example, a pair may select

Fig. 31. Golden eagle, adult attacking a prairie dog. After various sources.

a jackrabbit as their target, with the two birds flying separately and at different altitudes. One bird will distract the jackrabbit's attention while stooping toward it, while the other comes in to make the kill from another direction. Most cooperative hunting involves large prey, such as foxes or ungulates. Large prey animals are typically eaten where they are killed, but when feeding young at the nest. the limbs of young ungulates may be disarticulated and brought to the nest. It is unlikely that animals weighing more than about 4 or 5 pounds can be carried for any great distance. However, a bird-banding acquaintance of mine (who weighs slightly over 100 pounds), when once releasing a rehabilitated adult golden eagle in a stiff breeze, briefly lost control of its wings and the bird tried to take off. She found herself momentarily lifted a few inches above the ground, illustrating the terrific lift that can be generated by the eagle's broad wings, at least when aided by a strong breeze.

FERRUGINOUS HAWK

In many ways the ferruginous hawk (Fig. 32) is a golden eagle in miniature. In body mass adult ferruginous hawks weigh about one-third the amount of golden eagles, but the ferruginous hawk and the golden eagle share the same

Fig. 32. Ferruginous hawk, adult in flight with prairie dog.

fondness for open country with scattered elevations suitable for perching or nesting, especially in the form of rimrock areas, buttes, or rocky outcrops.

Being much smaller than golden eagles, the ferruginous hawk also has much smaller territorial and home range requirements. The home ranges of fourteen breeding males in Idaho have been judged in one study to average 2.3 square miles, as compared with an average of 3 square miles for seven males in Utah. Other home range estimates of 1.2 to 3.1 square miles have been made as well as a few much larger ones. These figures are roughly only 20 to 35 percent those of the home range estimates mentioned above for golden eagles. However, in eleven study areas scattered across the western United States, the mean distance between adjacent ferruginous hawk nests was 6.3 miles, according to a review survey by Marc Bechard and Josef Schmutz (1995). These inter-nest distance averages are considerably larger than the figures cited above for inter-nest distances of golden eagles in presumably good Wyoming habitat. No doubt prey densities influence nest spacing in ferruginous hawks just as they do in golden eagles, and the presence of competing red-tailed hawks and Swainson's hawks may also influence the local abundance levels and nest densities of ferruginous hawks.

Like golden eagles, ferruginous hawks often winter near prairie dog towns, and east of the continental divide they tend to concentrate on ground squirrels and prairie dogs as a reliable food source. West of the continental divide jackrabbits and cottontails are the primary prey. Collective prey data on more than 6,200 items identified in twenty studies and summarized by Bechard and Schmutz (1995) indicated that ground squirrels and prairie dogs accounted for 52 percent of 5,166 mammal remains, as compared with 24 percent represented by rabbits and hares. All told, ground squirrels and prairie dogs were present in 43.8 percent of all analyzed prey samples and composed an estimated 25.1 percent of calculated overall prey biomass. A relatively few prey remains of birds (822), of reptiles and amphibians (147), and of insects (68) were also found. Mammals the size of white-tailed jackrabbits (6 to 8 pounds) seem to represent the upper limit of their prey-killing selectivity, although these animals would be too heavy to be carried away in flight.

Ferruginous hawks and golden eagles forage in much the same way, sometimes by flying at considerable height while searching for prey on the ground and then swooping down silently and rapidly to grab its prey unawares. Unlike golden eagles, the hawks usually don't begin these searches at altitudes greater than 100 feet. Like golden eagles, ferruginous hawks also hunt prairie dogs and ground squirrels using "contour flights" as described earlier. Hovering almost stationary against the wind, with the bird positioning itself above an occupied burrow or potential prey, is sometimes used when wind conditions permit. A common and effective hunting method is to search visually for aboveground prey from an elevated perch and then take a low, short-distance flight of up to about 300 feet to surprise the quarry. When hunting pocket gophers and ground squirrels, the bird may wait patiently at a ground squirrel burrow or pocket gopher earthen pile until it detects soil movements or a nearby sound and then pounce on the ground and extract the unseen prey. There have also been reports of ferruginous hawks being attracted by gunfire to prairie dog towns, where they then feed on the newly shot animals.

During fall most ferruginous hawks migrate southward from the northern plains and gravitate to the southern dry grasslands, especially in the Staked Plains region of northwestern Texas. The highest winter concentrations occur near Amarillo and Dalhart, Texas, and west of the Davis Mountains near the Rio Grande Valley. From the Texas Panhandle the birds are abundant north to the Oklahoma Panhandle and to northeastern New Mexico, and south to a secondary peak around Midland and Odessa, Texas. Jackrabbits and prairie dogs are common throughout this part of Texas. A small winter population also regularly occurs in eastern Montana, around the Greycliff Prairie Dog Town Management Area near Big Timber.

D. L. Plumpton and D. E. Anderson (1998) reported that black-tailed prairie dogs are important winter prey species for ferruginous hawks in Colorado and that the hawks were most plentiful in areas where prairie dogs were most abundant. The birds hunted not only in larger areas of grassland habitat but also in some areas having high levels of human influence and habitat fragmentation, such as the Denver suburbs. D. J. Seery and D. J. Matiatos (2000) found that these hawks tracked prairie dog populations in the Rocky Mountain Arsenal National Wildlife Refuge. Similarly, J. F. Cully Jr. (1991) reported that ferruginous hawks were abundant in the Moreno Valley of northeastern New Mexico, where Gunnison's prairie dogs were abundant, but declined significantly when the prairie dog population suffered an outbreak of sylvatic plague. The population of golden eagles also declined then, but the red-tailed hawk population did not change significantly.

Ferruginous hawks have flexible nest-site requirements, but they often place their nests on elevated sites within large grassland areas. When trees are used for nesting, lone or peripheral ones are preferred over trees in densely wooded areas. Grassland areas seem to be preferred for nesting cover over cultivated ones, although nests close to roads or cultivated areas have been found to be most successful in Montana, probably because of the greater abundance of rodents in these ecological transitional habitats. However, nest success tends to be highest in areas of low human disturbance and where nests are placed in relatively remote locations. Many nests in the northern plains (e.g., Montana, South Dakota) are located at elevated sites with southern or western exposures, providing access to the prevailing southwestern winds of spring and summer and perhaps offering some early-season sunshine warmth for the sitting adult. Generally, ferruginous hawks avoid croplands and lands dominated by dense or tall sagebrush for nesting. In a western Kansas study by S. D. Roth and J. M. Marzluff (1989), 87 percent of ninety-nine nests were not in direct view of a prairie dog colony, but the majority of nests were within 5 miles of one. There most nest sites were surrounded by at least 50 percent rangeland and 25 to 50 percent cropland.

PRAIRIE FALCON

The prairie falcon (Fig. 33) has a quite different survival strategy than do the golden eagle and ferruginous hawk. It is a bird precisely designed by natural selection to have distinctly pointed rather than broadly rounded wings, and a fairly long and narrow tail that is rarely spread in the manner of a soaring eagle or buteo hawk. Its drag-reducing wing profile and body shape exhibit an airfoil more resembling that of a jet fighter than a cargo-transport plane, and their corresponding minimum of associated lift is compensated for by a maximum

Fig. 33. Prairie falcon, adult in flight. After various sources.

expenditure of energy. Thus it has adopted the same overall body design as have the peregrine, gyrfalcon, and merlin, with which high-octane speed is the essence of success and the leisurely aerial, circle-cutting flight patterns of eagles and buteo hawks are replaced by straight-arrow directness and simplicity.

Flight speeds of peregrines and prairie falcons are extremely hard to estimate accurately but are certainly faster than a prairie dog or ground squirrel can run. Prairie falcons use several hunting methods, including still-hunting from a perch, soaring at fairly high elevations, and prospecting at low levels. Although searching from perches may be the prairie falcon's most common method of hunting, low-level prospecting is its most characteristic method. The bird flies 10 to 20 feet above ground in a manner resembling a ground-hugging attack plane, dipping and rising as the land contours require, at a speed of perhaps 50 to 55 miles per hour. These flights often end in "straight attacks" from just above ground level. Higher attack speeds are no doubt possible during diving stoops, which are more typical of peregrines, but prairie falcons are generally thought to have somewhat slower flight speeds than peregrines.

Somewhat higher aerial approaches may end in low- or high-angle stoops of even greater velocity, and still higher-angle attacks may occur after takeoffs from a clifftop. If an attack is made and fails, a return flight and second attempt rarely occur, since the element of surprise has been lost. Instead, the bird is likely to disappear in the distance and begin searching anew. In one Oregon study 25 percent of 117 foraging attacks by prairie falcons were successful, but in Colorado only 13.5 percent of aerial attacks on horned larks resulted in prey captures, as summarized by Karen Steenhof (1998).

In most areas ground squirrels are the primary prey of prairie falcons, with such typical High Plains birds as western meadowlarks and horned larks being common secondary prey species. The size of prey taken by prairie falcons averages only about 4 ounces, with a reported maximum of up to about 4.6 pounds. As a result adult prairie dogs are substantially larger then the usual prey taken, but juvenile prairie dogs would be potential candidates. In Montana, white-tailed prairie dogs were reported as only a minor prey species in prey items examined by D. M. Becker (1979), with the remains of all mammals composing only 22.8 percent of the total sample, while those of birds made up the remainder (77.2 percent). However, P. A. McLaren and others reported white-tailed prairie dogs to make up 19 percent of prey samples they analyzed from Wyoming, with prairie dogs and ground squirrels the species' primary prey species (McLaren, Anderson, and Runde 1988). The prairie falcon was judged by these authors to be the most specialized local raptor as to its foods and foraging, at least as compared with the golden eagle, red-tailed hawk, and ferruginous hawk. In the study the estimated prey biomass of the prairie falcon was 98.4 percent mammalian and 1.6 percent avian.

Prairie falcons are monogamous, and pairs tend to nest in the same area year after year. The nest sites, or eyries, are usually on cliffs, escarpments, or steep slopes and most often have some kind of overhead protection. Two historic records of prairie falcons nesting in trees exist, but such behavior must be considered atypical. Although the same nest is usually used in subsequent years by a pair, a small percentage (22 percent in one survey of 161 birds) may change territories. However, about 80 percent of the pairs in the survey used the same territory at least two years in a row.

Prairie falcon home ranges are usually very large but vary regionally according to prey base sizes and prey distribution patterns. In Wyoming, Stanley Anderson and John Squires (1997) found that the home ranges of six prairie falcons averaged 43 square miles. In the Snake River Canyon area of Idaho the home ranges are much larger, averaging about 115 square miles. Generally, the birds with the largest home ranges have relatively poor reproductive success, since they must spend much energy getting to and from the nest with food. Winter home ranges are typically much smaller than summer ranges, as the birds can concentrate near rich food sources without being tied to a specific nest site.

Prairie falcons move to areas of high food resources during the colder months.

One of the densest concentrations in the High Plains is along the Nebraska–South Dakota border near the Fort Niobrara and Valentine National Wildlife Refuges. Another area of winter abundance is along the Arkansas River Valley near Great Bend, Kansas, not far from Cheyenne Bottoms Wildlife Management Area and Quivera National Wildlife Refuge. In northeastern Colorado winter concentrations are found in the Pawnee National Grassland and south-centrally along the San Luis Valley. In eastern New Mexico there is also a wintering concentration in the Kiowa National Grassland. All these areas have good numbers of wintering grassland birds and High Plains mammals.

OTHER HIGH PLAINS RAPTORS

The great horned owl is such a widespread species and such an extremely opportunistic forager that one might expect prairie dogs to appear often on its list of common prey. Florence Bailey (1928) listed prairie dogs among the documented prey of great horned owls in New Mexico, and black-tailed prairie dogs were also mentioned as being among the many prey species of great horned owls in a recent comprehensive review (Houston, Smith, and Rohner 1998). Although great horned owls forage mostly during the evening and early morning when prairie dogs are likely to be underground, the owls may also occasionally forage during daytime hours when trying to gather food for their young; this is when prairie dogs may be most vulnerable.

Red-tailed hawks are large enough to pose a significant threat to prairie dogs, as they regularly take animals as large as black-tailed jackrabbits. However, like great horned owls, they are so opportunistic in their prey selection that they probably kill prairie dogs only infrequently relative to more common, more widespread, and perhaps less wary prey. Mammals up to about 4.5 pounds do predominate in the diet of red-tailed hawks, and in the western and northern parts of North America these mammals mostly consist of jackrabbits, snowshoe hares, and ground squirrels. In an Alberta study (Luttich, Keith, and Stephenson 1971), mammals comprised sixty-two of the prey items counted, with Richardson's ground squirrel alone making up 32 percent and black-tailed jackrabbits another 17 percent.

Swainson's hawks are somewhat smaller than red-tails and concentrate on correspondingly smaller prey, so their exploitation of prairie dogs is likely to be even more infrequent. In the northern plains regions, such as North Dakota and Saskatchewan, 35 to 40 percent of the prey items during the summer months are typically ground squirrels, especially Richardson's ground squirrel, and 25 to 55 percent consists of other rodents, including many small mice and voles, based on a comprehensive review (England, Bechard, and Houston 1997).

Snowy owls and rough-legged hawks often visit prairie dog colonies during their winter visits, as do bald eagles at times. There is no evidence that some other larger prairie raptors, such as the short-eared and the northern harrier, pose any real threat to prairie dogs, as they both concentrate on smaller rodents. And even smaller raptors, such as American kestrels and screech-owls, may not pose a threat even to juvenile prairie dogs and perhaps are attracted to them to exploit the relative abundance of small rodents typical of prairie dog colonies. Like most members of the prairie dog–grassland ecosystem, they can only enchant us and make us wish we had more time to enjoy them and try to understand their places in this ecological web of life.

This curious world which we inhabit is more wonderful than it is convenient, more beautiful than it is useful; it is more to be admired than used. – *Henry David Thoreau*

8

The Varmint and Predator Wars

It's Finally Almost Quiet on the Western Front

Legislation passed by the U.S. Senate in 1930 called for "the destruction of all mountain lions, wolves, coyotes, bobcats, gophers, ground squirrels, jackrabbits and other animals injurious to agriculture, horticulture, forestry, husbandry, game or domestic animals, or that carried disease."

As a result of Federal and State cooperation, satisfactory progress has been made in the control of the larger predatory animals with the exception of the coyote, which has proved to be the archpredator of them all. – *E. A. Goldman, senior biologist, Bureau of Biological Survey (1930)*

In July 1915, at the behest of powerful cattle interests, the U.S. government took on the job of controlling predators throughout the United States. Between 1911 and 1914 the cattle ranchers had repeatedly applied to the U.S. Forest Service to control prairie dogs and predators on both their own and Forest Service–managed lands that they were grazing at cut-rate prices. However, the Forest Service, a branch of the Department of Agriculture, which had been established in 1862, demurred. Instead, they passed the assignment on to the U.S. Biological Survey, which had grown out of the Department of Agriculture's Division of Economic Ornithology and Mammalogy in the 1880s. C. Hart Merriam, the Survey's new director, took on the challenge with a vengeance. By 1919 the total acreage of prairie dogs in the West was estimated to have been reduced from 992 million acres in 1900 to about 100 million acres (Nelson 1919). Between 1915 and 1918 federal agents not only poisoned uncountable millions of prairie dogs but also killed 70,713 large predators, including 60,473 coyotes, 8,094 bobcats, and 1,829 wolves (Schlebecker 1963).

Between 1915 and 1947 at least 1,884,897 coyotes were destroyed by government agents, primarily in the western stock-growing states of Texas, Idaho, Wyoming, Oregon, Utah, Nevada, California, and Colorado. The control agents used various techniques, including rifles, traps, strychnine, and cyanide. In addition to accounting for nearly 2 million coyotes in thirty-two years (averaging more than 60,000 a year, or nearly twenty deaths a day), they tallied over 30,000 wolves and uncountable numbers of other carnivores, including badgers, skunks, and foxes (Dobie 1949).

A separate agency known as Predator and Rodent Control (PARC), a division of the Department of Agriculture, was established in 1929 to deal with other predators in the West, after the wolves had finally been effectively controlled. In 1931 the Animal Damage Control Act was passed. PARC was renamed Animal Damage Control (ADC) in 1939 and was assigned to the newly established U.S. Fish and Wildlife Service within the Department of Interior. This agency's activities flourished in the 1940s, especially after World War II.

During World War II, sodium monofluoroacetate, or Compound 1080, was discovered. It was found to be extremely lethal to canines in almost microscopic quantities but required up to eight hours to cause death, during which the animal was in agonizing pain. Compound 1080 had been developed as a military weapon, with plans to induce mass poisoning among the Japanese civilian population by introducing it into their water supplies. According to Edward Raventon (1994), government trappers in the Black Hills pursuing coyotes would shoot a wild range horse, pack it up, and set the pieces out on the prairies after injecting them with 1080. A day or two later they would inspect each piece to inventory the circle of death radiating out from it, which included not only coyotes but also skunks, foxes, bobcats, eagles, magpies, and crows. Starting in the 1960s thallium was also adopted, a tasteless poison so deadly that agents had to wear masks when handling it, and that rapidly kills almost every species of animal that happens to taste the bait. Thallium became a valuable tool in the arsenal against coyotes after 1080 was banned.

For example, during one year in the 1960s, Predator and Rodent Control employees eliminated 89,653 coyotes, 24,273 foxes, 6,941 badgers, 2,779 wolves, 842 bears, and 194 mountain lions. As summarized by Hope Ryden (1979), all of this was accomplished on a national budget allotment of only about $7 million, representing only $50 per individual predator disposed of. Unfortunately, as indexed to subsequent inflation, $7 million is only slightly more than is now spent for federal predator control activities among the states listed in Table 7, exclusive of the other forty-five states in the Union.

COYOTES

By 1970 the United States was spending about $8 million annually to kill about forty thousand coyotes per year, at an average cost then to taxpayers of about $200 per coyote. However, in 1972, the use of poison on public lands was outlawed, and in 1973 further restrictions were enacted by the Environmental Pro-

Table 7. Coyotes reported killed by sport hunters and taken in USDA's wildlife services program during 2000–2001 in five High Plains states

STATE	SPORT-HUNTER KILLED (YEAR)	USDA WILDLIFE SERVICES PROGRAM TAKEN	TOTAL KILLED
Colorado	21,350 (2000)	3,526	24,876
Nebraska	25,221 (2000–2001) [a]	3,234	28,455
North Dakota	2,176 (2000)	2,578	4,754
South Dakota	6,391 (2000–2001)	Not reported	6,391+
Wyoming	1,254 (2000)	7,857	13,863
All states	191,482 (34 states)	88,665 (32 states)	280,147 [b]

Source: Data from USDA's Wildlife Services Program web site (2003), for FY 2001. State hunter-killed data is for indicated twelve-month periods, as reported to USDA. Private trapper take is not included in data.
[a] The 1998 estimate for combined hunter and trapper kill in Nebraska was 38,896 animals.
[b] National estimate for current annual combined hunter and trapper kill is ca. 350,000 animals.

tection Act. Counting both professional and amateur trappers, the number of coyotes taken per year during the 1960s and 1970s is not available but is certainly much higher than that reported in Table 12 (in Chapter 10). Hope Ryden (1979) stated that in 1977 the total reported take of coyotes by trappers alone was 303,932 pelts.

After negative national publicity in the l960s and 1970s, such as that described for coyotes by Hope Ryden above, Animal Damage Control was transferred in 1986 to become a part of the U.S. Department of Agriculture's bureaucracy innocuously known as APHIS (Animal and Public Health Inspection Services). In recent years Animal Damage Control has euphemistically been renamed Wildlife Services.

Interestingly, the number of coyotes taken annually by the Animal Damage Control agents in the 1960s was almost identical to that reported by Wildlife Services for 2001, nearly ninety thousand animals. Of the thirty-nine states reporting apparent coyote population trends to Wildlife Services in 2001, twenty-one states reported stable populations, seventeen judged that populations were increasing, and one state (North Dakota) reported decreasing populations. Thus, despite forty years of diligent efforts by state and federal agencies, ranchers, amateur trappers, and rifle-toting ranchers; despite massive increases in state and federal funding; and despite improved methods of killing, the national coyote population survives and even flourishes. As Adolph Murie (1940) noted, although 4,325 coyotes were killed in Yellowstone National Park between 1907 and 1935 (when their hunting was finally banned), their population at the end of this period was greater than it had been at the start of the twentieth century.

Three decades of coyote killing had served only to make the surviving animals even smarter.

Coyotes were taken in fiscal year 2001 by Wildlife Services most frequently through aerial hunting (40.4 percent), followed by use of sodium cyanide ("M-44") cartridges (18.1 percent), attraction to predator calls (9.6 percent), and leg-hold traps (7.8 percent). The total animals taken during fiscal year 2001 totaled over 1.7 million animals of more than 200 species (U.S. Department of Agriculture, Wildlife Services 2003), including nearly 90,000 coyotes, 4,600 red and gray foxes, 650 badgers, 390 bears, 390 mountain lions, 118 wolves, 24 kit foxes, and even 14 nationally threatened swift foxes (oops). Wildlife Services activities in fiscal year 2001 were funded in the amount of $1,835,000 in Wyoming, $1,442,000 in South Dakota, $1,306,000 in North Dakota, $1,215,000 in Colorado, and $774,000 in Nebraska. In Nebraska some 3,200 coyotes were taken during 2001, which, if considering only coyote control, would also have cost about $400 for each animal taken, not counting all the other animals disposed of by Wildlife Services at no additional charge to taxpayers. In North Dakota agents eliminated 2,071 coyotes and 1,026 beavers during a single recent year. This was achieved at an overall state and federal cost of $810,000 and during a year when some $126,000 in livestock damage was reported in the state. That works out to somewhat less than $400 per coyote carcass, or twice the cost during 1971 and eight times the cost during the 1960s. It also represents $6.42 spent in response to every dollar of livestock damage reported, which represents typical federal cost-to-benefit mathematics.

Similarly, in Arizona $157,603 was spent to control coyotes that had been deemed to have caused $42,225 worth of livestock damage, a slightly better cost-to-benefit ratio of 3.7 to 1. And in New Mexico agents killed twenty-eight coyotes as well as some skunks and foxes and a few nonpredators, such as porcupines and deer, during five hundred agent-hours effort, in response to a rancher claiming to have lost a lamb worth $83 (Grady 1994). Assuming at least $20 per hour was paid the agents, the total costs exclusive of travel and supplies must have been at least $10,000, a cost-to-benefit ratio of 120 to 1.

Running coyotes to death from exhaustion is one favored method of coyote haters for making sure the animals do not die quickly or easily, but it is probably better than dying from poisoning by cyanide, thallium, or Compound 1080. Other more diabolical methods of torture have included wiring a coyote's mouth shut and releasing it to starve; sawing off its lower jaw; tying a gunnysack soaked with kerosene around a coyote's body and setting it ablaze; and leaving an coyote in a leg trap until it dies of starvation or massive infections of the injured leg. A wonderfully ironic story was recounted by Wayne Grader, who told of a rancher in Wyoming who captured a coyote that he thought might be attacking his sheep. Rather than killing it directly, he tied a stick of dynamite to the animal's body, lit it, and released the coyote. The terrified animal ran quickly to the rancher's new pickup truck and took refuge underneath it.

The cattle and sheep ranchers have become an even stronger political force in recent years, and what they may lack in ecological knowledge is more than made up for in money and political influence. There are is now the National Cattlemen's Association, with fifty state chapters; the Beef Industry Council, with forty state chapters; thirty different National Breed Associations; and other national and state groups associated with the cattle industry that collectively take in $100 million per year in dues and tax checkoffs. Obviously, with that kind of money they always have the attentive ears of the Department of Agriculture. Then there are the large numbers of equally influential National Rife Association members and other gun owners who may not care about protecting sheep or cattle but simply like to use coyotes for target practice, knowing they are exhibiting their constitutional rights to bear arms and shoot anything that is not legally protected (see Table 7).

One wonders whether someone in your department has gone mad from a personal hatred of predators. . . . We wonder what kind of misfits may be perpetrating this campaign. – *Letter of Dr. Raymond Bock, Pima Medical Society, Pima, Arizona, to the U.S. Department of Interior relative to government-sponsored predator control efforts*

Some men consider calling up coyotes and shooting them sport. – *J. Frank Dobie,* Voice of the Coyote *(1949)*

PRAIRIE DOG CONTROL

If the stock farmers of the American West have traditionally detested coyotes, then prairie dogs have similarly been considered as beneath contempt. An early phase in the alliance between the federal government and stockmen interests occurred in 1902, when C. Hart Merriam estimated (without any experimental justification) that 256 prairie dogs might consume as much grass as a one-thousand-pound steer and that thirty-two of them would eat as much as one sheep; he further stated that prairie dogs contributed to a 75 percent reduction in rangeland productivity in the High Plains, again without providing scientific proof. Merriam's statements provided the official cachet for eliminating prairie dogs in the American West. They were only the first salvo in the government's subsequent century-long war on prairie dogs. Merriam was soon supported in his claims as to the ominous if not devastating effects of prairie dogs on agricultural and ranching interests by other Department of Agriculture employees, such as Vernon Bailey and W. R. Bell (1919), again without scientific proof.

The figures produced by C. Hart Merriam have since been strongly disputed by biologists. John Hoogland (1994), for example, has pointed out that prairie dogs often eat plants not consumed by livestock, although both prefer the same plants. Prairie dogs also tend to colonize already overgrazed areas and contribute to forage quality for ungulates by effecting mineral nutrient mixing of the soil substrate and soil aeration through their digging activities. As a result, estimates of unfavorable competitive effects between prairie dogs and cattle have historically been greatly overestimated.

About twenty years after Merriam made his assessments, two other Department of Agriculture researchers, W. P. Taylor and J. V. G. Loftfield, further asserted (1924) that the Gunnison's prairie dog is "one of the most injurious rodents of the Southwest and plains region." The federal government backed up the strong words of Merriam, Bailey, Bell, and subsequent advocates with action. Between 1915 and 1920 about 47 million acres of prairie dog colonies were been poisoned in six western states (Bell 1921). Between 1912 and 1923 nearly 45 million acres of land in Colorado alone were treated with poison bait for eliminating both prairie dogs and ground squirrels. Typically, both cyanide and strychnine were coated on wheat bait, which of course killed all species of grain-eating mammals, birds, and all other animals that happened to consume it. By 1922 government agents had already eliminated about 90 percent of the prairie dogs living in the Texas Panhandle by poisoning about a million acres of colonies.

By the late 1990s the government had done its job well. Close to 99 percent of all the black-tailed prairie dog acres in North America had been eliminated in little more than a century (Table 8). In Arizona the kill of black-tailed prairie dogs was a perfect 100 percent, eliminating that species of prairie dog from the state. These efforts also probably reduced the ranges and population sizes of the ferruginous hawk and burrowing owl (Glinski 1998) as well as such other prairie dog associates as the golden eagle.

The white-tailed, Gunnison's, and Utah prairie dogs fared no better at the hands of the government and of private landowners during the 1900s. Very little information exists as to the presettlement populations of any of these species, but Craig Knowles (2002) has summarized what is known about the first two. For example, one white-tailed prairie dog population in northwestern Wyoming that covered an estimated 200,000 acres in 1915 was reduced to less than 1,000 acres by 2002, a 99 percent reduction. In New Mexico, some 11,150,000 acres of Gunnison's and black-tailed prairie dogs were poisoned between 1917 and 1932.

Table 8. Estimated black-tailed prairie dog acreages, 1870 and 1998

| | TOTAL ACTIVE ACREAGE | | |
STATE	1870	1998	DECLINE (%)
Arizona	650,000	0	100.0
Colorado	7,000,000	44,000	99.4
Kansas	2,500,000	36,000	98.6
Montana	6,000,000	65,000	98.9
Nebraska	6,000,000	60,000	99.0
New Mexico	2,000,000	15,000	99.9
North Dakota	2,000,000	15,000	99.3
Oklahoma	950,000	8,500	99.1
South Dakota	1,757,000	244,500	86.0
Texas	56,833,000	22,500	99.9
Wyoming	16,000,000	125,000	99.2
Total U.S.	116,000,000	635,000	99.5
Saskatchewan	NA	1,048	NA
Chihuahua	NA	136,487	NA
GRAND TOTAL		772,535	

Note: NA = not available.

Sources: State estimates are based on National Wildlife Federation petition (1998). These often are less than
the usually much larger estimates reported by state agencies, which commonly include inactive acreages.
These larger figures and some alternate recent estimates were provided by Rosmarino (2003). The Saskatchewan estimate (for 2000) is also from Rosmarino (2003), and the Chihuahua estimate (for the early
1990s) is from Miller (1996).

* Prairie dogs were first discovered in Canada in 1937. Nearly all Canadian colonies (96%) now occur
within Grasslands National Park, Saskatchewan.

From 1933 through 1943 in New Mexico, 8,550,000 acres of Gunnison's prairie
dogs were poisoned. From these and related data, Knowles concluded that there
may have been more than 4.5 million acres of Gunnison's prairie dogs in New
Mexico as of 1919, distributed over about 15.3 percent of the total state landscape.
By 1982 this acreage had been reduced to an estimated 75,000 acres, representing
a nearly 98 percent reduction. The other states' do not seem to have enough
historical information to provide comparable estimates of population declines.

Recent poisoning campaigns in Nebraska, South Dakota, and Wyoming provide some measure of how private individuals, state agencies, and federal Department of Agriculture agents have been proceeding with prairie dog control
efforts during the late 1990s and early 2000s (Table 9). In Nebraska it has been
found that a 70 to 80 percent control of prairie dogs may be achieved with zinc
phosphide–treated oats or zinc phosphide pellets, for only $7.57 to $7.75 per

Table 9. Recent prairie dog poisoning statistics for three Great Plains states

STATE	METHOD USED	YEAR	AMOUNT USED	ACRES TREATED
Nebraska	Aluminum phosphide gas (tablets) [a]	1996	>24,109	
		1997	32,020	
		1998	23,100	
		1999	31,659	
		2000	20,217	
		TOTAL	131,105	3,933
	Zinc phosphide bait (lbs.) [b]	1996	15,258	
		1997	400	
		1998	549	
		1999	20,453	
		2000	16,811	
		2001	12,600	
		TOTAL	50,813	55,386 (est.)
	Gas cartridges	1996	0	
		1997	0	
		1998	100	
		1999	117	
		2000	158	
		TOTAL	375	
South Dakota	Aluminum phosphide gas (tablets)	FY 2000	500	
	Zinc phosphide baits (lbs.)	FY 2000	29,825	32,509 (est.)
		FY 2001	28,825	31,419 (est.)
Wyoming	Zluminum phosphide gas (tablets)	2000–2001	713	
	Zinc phosphide bait (lbs.)	2000–2001	63,007	68,677 (est.)

Sources: Data primarily from Rosmarino (2003), and Wildlife Services web site (2003), including USDA's Wildlife Services program (Nebraska), Department of Agriculture data (South Dakota), and reports of private sales (Wyoming).

[a] A fumigant releasing hydrogen phosphide gas. Typical prairie dog application rate is two 3-gram tablets per burrow (Moline and Demarais 1988), or about one tablet per 0.03 acre.

[b] Baits are typically 4-gram pellets, containing 2.0% zinc phosphide. Typical application rate is 1.09 acres/pound, which is the basis for the estimated total acres treated (Rosmarino 2003; Deisch, Uresk, and Linder, 1989).

acre treated (Hyngstrom and McDonald 1989). Assuming a prairie dog density of ten animals per acre, that represents a cost of only a dollar per carcass.

Zinc phosphide is a notably effective killer of a wide range of rodents, including rats, voles, ground squirrels, cotton rats, and prairie dogs. It has also been used to control jackrabbits, woodrats, pocket gophers, nutria, muskrats, and moles. Desert kit foxes (and presumably also swift foxes) have mean lethal doses that require about 5 to 7.5 times the dosage of rodents, but domestic fowl, such as chickens and geese, appear to be about as sensitive to zinc phosphide as are rodents. Wild seed-eating birds presumably are similarly sensitive. Dogs, cats, and pigs are known to have consumed poisoned rodents and died as well. However, eagles, vultures, red and gray foxes, and coyotes similarly have eaten poisoned rodents without reported mortalities. Over about fifty years some twenty-six human deaths are known to have been caused by zinc phosphide, most of which were suicides, plus a few murders (Marsh 1989).

In recent years professional animal control agents have received more assistance from "recreational" shooters, for whom the ability to disintegrate a prairie dog at a distance of several hundred yards with a scope-mounted, high-powered rifle is considered a special accomplishment (Table 10). Few statistics on the numbers of animals annually taken by this method exist, as no record-keeping data are required, unless one might be hunting on a national grassland or an Indian reservation where such records may be kept. For example, at Fort Belknap Indian Reservation in Montana, the income for sales of shooting permits on a per-acre basis has exceeded what is collected by the U.S. Bureau of Land Management for grazing on public lands. As of the mid-1990s two of the central and northern plains States (Montana and Wyoming) still had no licensing requirements for regulating prairie dog shooting. Kansas, Colorado, and South Dakota required a small-game hunting license to shoot prairie dogs, whereas in Nebraska and North Dakota a small-game license to kill prairie dogs was needed only for nonresidents of the state. None of the states had a specific hunting season or bag limits on numbers taken (FaunaWest 1998); nor did New Mexico, Oklahoma, or Texas. Although hard data are difficult to obtain for such reasons, it has been estimated that in 1999, 7,100 shooters killed 300,000 prairie dogs in Nebraska, and in Colorado 3,400 shooters killed more than 250,000 prairie dogs. In South Dakota, about 1.52 million prairie dogs were killed by recreational shooters in 2001, in addition to unknown numbers shot on tribal lands. An estimated 16,011 hunters were involved, representing an average of 95 animals killed per person (South Dakota Department of Agriculture).

RECENT BLACK-TAILED PRAIRIE DOG CONSERVATION AND MANAGEMENT EFFORTS

With the recent threats by the U.S. Fish and Wildlife Service to the eleven states having black-tailed prairie dog populations to produce individual state conservation measures or face the possibility of having the species placed on the threatened list and receive federal protection, changes in state hunting regu-

Table 10. Some estimated annual kills of prairie dogs by recreational shooting

STATE	YEAR	HUNTERS	ESTIMATED KILLS
Colorado (all species)	2000–2001	3,369	229,502
Kansas (black-tailed)	2001	NA	167,000
Nebraska (black-tailed)	1998	NA	301,000
	1999	NA	356,000
South Dakota (black-tailed)	1999	NA	1,435,437
	2000	NA	1,246,842

Note: NA = not available.
Source: Data from Rosmarino 2003.

lations, restrictions on poisoning, and habitat protection plans are finally likely to occur, at least on publicly owned lands.

As of 2002 the Arizona Game and Fish Commission was considering the reintroduction of black-tailed prairie dogs and continued to ban the shooting of this entirely extirpated species on its nonexistent colonies. In New Mexico, no state plan for managing the black-tailed prairie dog had been finalized as of 2002, and a draft plan under consideration failed to restrict shooting, poisoning, or habitat destruction on private or public lands. Data collection and conservation measures on private lands remain voluntary.

Also as of 2002 Colorado had eliminated the shooting of black-tailed prairie dogs on public lands but had not yet regulated the shooting of Gunnison's and white-tailed prairie dogs. Nor had the shooting of black-tailed prairie dogs on private lands for control purposes been restricted in that state, and poisoning on state-owned lands is permitted by state statute. In 2001–2002 approximately 452,000 black-tailed prairie dogs were killed by 3,703 recreational shooters in the state, as compared with 229,502 prairie dogs shot by hunters the previous season. In 2003 Colorado produced a draft Conservation Plan for Grassland Species in Colorado that recognized the importance of prairie dogs in the grassland ecosystem (Colorado Division of Wildlife 2003), and proposed accepting a minimum target of 255,733 acres of occupied habitat, as proposed by the Black-tailed Prairie Dog Conservation Team (Luce 2003). This draft document estimated that as of 2002 Colorado supported 631,102 acres of prairie dogs, a figure that Rosmarino, Robertson, and Crawford (2003) regarded as most likely "inflated."

Montana legislators passed a bill in 2001 that recognized the black-tailed prairie dog as a "species in need of management," although it is still considered a pest species by the Montana Department of Agriculture, and so population controls on private lands continue. The state lands department also is allowed to poison and shoot prairie dogs through agreements with the Montana Department of Fish, Wildlife and Parks. Shooting on federally owned lands was terminated during most of the spring breeding season, and year-round protection was provided for one rather small (20,000-acre) area.

North Dakota has so far refused to sign any agreement regarding prairie dogs conservation, other than to try to maintain the colony acres (about 20,000) found during recent surveys, including Indian reservation lands over which the state has no direct control.

South Dakota, which supports one of the largest prairie dog populations in the High Plains, agreed in 2001 to change the species' "pest" status to that of "species of management concern." The South Dakota Game Fish and Parks Commission also proposed the prohibition of prairie dog shooting on public lands during the spring breeding season, although this restriction does not apply to private or tribal lands. If the state prairie dog population seriously declines, the state's Department of Agriculture may cease the manufacture and sale of zinc phosphide poison and, with further population losses, may prohibit the use of this and other pesticides for controlling prairie dogs. In February 2004 the U.S. Forest Service lifted its ban on prairie dog poisoning in the Dakotas, Nebraska, and Wyoming. The decision affected the Nebraska National Forest and Oglala National Grassland, the Buffalo Gap and Fort Pierre national grasslands of South Dakota, all five of the national grasslands of North and South Dakota, and Wyoming's Thunder Basin National Grassland.

In 2002 the Nebraska Game and Parks Commission unanimously defeated a resolution calling for a minor change in the black-tailed prairie dog's wholly unprotected status (to be upgraded to a "species in need of conservation") and possibly adopting a state management plan for prairie dogs, which have been classified as an "unprotected nongame species." No season limits, restrictions on seasonal hunting periods, or bag limits exist. Residents need not even possess a small-game hunting permit to shoot prairie dogs, but nonresidents must at least possess such a permit. Hunting on federal lands – such as the Oglala National Grassland, one of the state's few federally owned lands with the potential for supporting a classic prairie dog–shortgrass ecosystem – remains unrestricted. This situation reflects the Forest Service's unwillingness to interfere with local cattle-grazing interests, as exemplified by the cozy relationship existing between the Forest Service and the Sugar Loaf Grazing Association. The state's governor, Mike Johanns, long a critic of conservation activities, supported the commission's actions in regard to prairie dogs, noting that land used by prairie dogs becomes "useless for feeding cattle." He didn't add that land used only for feeding cattle eventually becomes useless for everything.

In 2002 the Kansas Department of Parks and Wildlife established a prairie dog conservation and management plan, classifying the species as "wildlife" and recommending that the state law requiring mandatory eradication of prairie dogs by landowners be changed to a voluntary statement. In Kansas a hunting license is required to kill prairie dogs, but no seasonal restrictions or bag limits are in effect.

Oklahoma had not produced a draft of a state prairie dog management plan as of late 2002, and the vast majority of prairie dog towns in the state are on land receiving no protection of any kind.

The Texas Parks and Wildlife Department has not proceeded beyond a draft of a prairie dog management plan and in 1998 tried to avoid a federal listing of the species by proposing an interstate conservation agreement by the affected states. Texas's own working group for such a plan had not adopted a final version as of early 2003. However, their draft plan calls for a statewide inventory of prairie dogs and an educational program that may increase public tolerance for prairie dogs on private and public lands. In Texas the species is treated as a "nongame" species by Texas Parks and Wildlife and as a "pest" species by the Texas Health and Safety Code.

In May 2004 the U.S. Fish and Wildlife Service announced that it had not yet updated its 2002 finding with regard to the black-tailed prairie dog, in which a listing proposal for this species was considered as warranted but was precluded by higher priorities, and for which a relatively low level listing priority number of 8 (with a priority rank of 1 being the highest) had been assigned. [See page 145 for updated status.]

Prairie dogs can be devastating to ranchland. They can effectively destroy a ranch by burrowing, chewing up pasture, and leaving numerous holes spread across a field. Ground invaded by prairie dogs is useless for feeding cattle and for that reason is significantly less valuable should a producer decide to sell. – *Mike Johanns, governor of Nebraska, letter of August 27, 2001*

If we can extend the idea of community to include the lowliest of creatures, we will be closer to finding a pathway to empathy and tolerance. If we cannot accommodate them, the shadow we see on our own home ground will be a forecast of our extended winter of the soul. – *Terry Tempest Williams, "In the Shadow of Extinction,"* New York Times, *February 2, 2003*

9

Taming the Great American Desert

Hardscrabble Times at the Fringes of Nowhere

Grass no good upside down. – *Anonymous Pawnee chief, on first seeing prairie lands of the Great Plains being plowed*

The settlement of the High Plains was given a major nineteenth-century impetus by the Homestead Act of 1862, awarding 160-acre allotments to adventurous homesteaders who were willing and able to live for at least three years on the land they had acquired by simply paying a low registration fee. The Homestead Act had been preceded by the Pre-emption Act of 1842, which allowed squatters to buy (for $1.25 per acre) 160-acre parcels of land that they had occupied for at least six months (a period that was extended to at least fourteen months in 1891). The Homestead Act opened up vast tracts of land in the drier parts of the West, just as railroads were penetrating into these previously almost inaccessible regions, and was opposed by cattle interests as an invasion of their already heavily used (but not actually owned) grazing lands. The Timber Culture Act of 1872 provided additional 160-acre allotments to persons who planted part of their homestead to trees. The Kincaid Act of 1904, providing 640-acre rather than 160-acre homestead allotments, originally applied only to a relatively few arid and sandy counties in western Nebraska but was later expanded to include eleven additional states through the Enlarged Homestead Act of 1909 and the Three-Year Homestead Act of 1912. In the nine affected states of the High Plains region, homesteading resulted in the settlement of about 160 million acres, an area almost as large as Texas (see Table 12 in Chapter 10).

No federal homesteading occurred in Texas, which had entered the Union in 1845, almost twenty years prior to the passage of the Homestead Act. Having previously been an independent republic, Texas retained all of its public lands, so there were no federal disbursements of its lands under the Homestead Act. It later developed its own version of the Homestead Act, which allowed for homesteading parcels of 4,470 acres. These large land allotments, which took into account the aridity of the land and the ranching-based economy, provided one of the state's early examples of "thinking big." The state also sold about 5 million acres of its land to provide for education of the state's children.

The Desert Land Act of 1877 gave 640-acre allotments in the more arid western states, all for the bargain-basement price of $1.25 per acre, provided the landholder agreed to begin irrigating the land within three years of occupancy. This three-year provision was an invitation to fraud. Ranchers gained entry to these lands by providing minimal down payments of 25 cents per acre, grazed the land until it was bare, and let their claims lapse. The similar Timber and Stone Act of 1878 did not affect High Plains settlement, as it was limited in its coverage to the Pacific Coast states and Nevada. However, it gave away the rich forest lands of the Pacific Northwest to speculators at comparable rock-bottom prices. By the end of the 1800s there were 40 million cattle and 40 million sheep grazing on publicly owned lands in the West, and 700 million acres of publicly owned lands had already been ruined by overgrazing (Lyman and Metzer 1998).

After the invention of barbed wire in 1874, farmers were finally able to keep range cattle off their homesteaded lands and away from their existing water sources and their windmill-based water supplies, helping to start the rancher-farmer wars of the late 1800s. With the advent of barbed wire fencing, the era of open ranges and long cattle drives from Texas to the railroad market terminals of central Kansas and western Nebraska began to draw to a close. The Carey Act of 1894 provided that the federal government would give to the individual western states the remaining public lands of their arid regions, the proceeds of which were to be used for in-state reclamation purposes. Under this act Montana, Wyoming, and Colorado each received 2 million acres to be "reclaimed" through state-run irrigation projects (Webb 1931). The era of large-scale irrigation of the High Plains, and the associated diversions of surface waters as well as the tapping of the vast Ogallala aquifer below them, was about to begin. A century later the aquifer had begun to dry out, starting at the state's southern end, with desertification occurring in Texas, and the Kansas Water Office has calculated that 75 percent of its irrigated acreage will be lost by 2020. By that time almost 3 million acres of irrigated land will have to be abandoned or reverted to dryland farming (Callenbach 2000).

The railroad companies encouraged the settling of the High Plains, even describing them in Eden-like terms to potential farmers, as a place to grow enough wheat to feed the booming American population. As the railroads pushed farther into the western plains – with the Chicago, Milwaukee, St. Paul, and Pacific reaching Butte, Montana, in 1907 and Seattle in 1909 – the farmers followed. Like the advent of barbed wire, the railroads helped to subdivide the West into ever smaller units, converting the economy of the High Plains from one based largely on open-country ranching to a dryland farming culture. The overall population of the Plains states increased 37.6 percent between 1900 and 1910, with the six northern states averaging a 27 percent increase and the four southern states increasing 46 percent. The number of farms in the ten Great Plains states also increased by 37 percent during that same period. Largely through immigration the entire U.S. population was rapidly growing, rising 7.4 percent between 1906 and 1910. During the 1900–1910 decade, the percentage of urbanites in the United States increased from 40 to 45.8 percent, as the eastern cities filled with recent arrivals hoping to fulfill their dreams in America (Schlebecker 1963). As the human population and goods requirements of America grew, its native grasslands disappeared under the plow, with the rich tallgrass prairies disappearing first and being replaced by corn and small grain crops. Next the mixed-grass prairies began to disappear, with dryland agriculture eventually taking between two-thirds and three-fourths of the total. Finally the shortgrass prairies and shortgrass-shrublands of the High Plains were replaced by irrigated farmlands or destroyed by overgrazing (Table 11).

Wheat production on the plains was of course directly controlled by annual rainfall variations, with periodic droughts striking the Great Plains in 1909–11, 1914, and again on the southern plains in 1917. The postwar depression of 1919–22 was supplemented by grasshopper outbreaks. Massive, new government-sponsored efforts at killing prairie dogs, other rodents, and predators of all sizes were also undertaken on the Great Plains. The northern plains were afflicted in 1919 and 1920 with the same drought conditions that had hit the southern plains a few years earlier. And, after a few good years from 1926 to 1928, the Great Depression hit in 1929. Local southern droughts began in 1929 and spread north the following year. By 1933 the entire Great Plains were enduring near-drought conditions, accompanied by plagues of grasshoppers. Between 1930 and 1935 some 150,000 people abandoned the Plains, nearly all of them from the northern plains states from Montana and North Dakota to Kansas. However, the population of the Great Plains states collectively continued to grow slowly, but at a rate of only about two-thirds that of the country as a whole. As farmers left the plains, cattle ranchers tended to move in to replace them, but as confined ranch or pasture farmers rather than large-scale, open-range ranchers (Schlebecker 1963).

The peak of the drought period occurred in 1934, but 1936 was nearly as bad, with dust storms during the unbearably hot summers followed by some of the coldest winters in U.S. history. This writer was born during 1931 in a tiny North Dakota village and vividly remembers both the heat and the cold of those terrible near-poverty years. Grasshoppers appeared each summer with a vengeance, as did jackrabbits, prairie dogs, and ground squirrels, leaving very little if anything for the embattled small-grain farmers to harvest. The government's unrelenting predator control efforts on wolves, coyotes, and foxes had clearly worked all too well and had resulted in exploding populations of rodents and hares. Between 1936 and 1939 the human populations of the dryland, wheat-growing states of the Dakotas, Nebraska, and Kansas actually declined (Schlebecker 1963).

With changes in land use and the economics of farming and ranching on the High Plains during the 1900s, individual properties became larger and landowners ever fewer. By the end of the twentieth century only 5 percent of U.S. landowners owned 50 percent of all the farmlands, and nine out of ten landowners were nonfarmers. Ninety-eight percent of the landowners were white, and two-thirds were more than sixty years old. Under the 1996 Freedom to Farm Act, federal farm-aid subsidies reached $5.7 billion in 1996, $6 billion in 1997, and $7.4 billion in 1998. In a traditional farm state such as North Dakota, for example, farm subsidies provided about half a billion dollars annually to the state, representing 80 percent of total farm income. Grazing-fee subsidies by the 1990s totaled $5 million for the ten westernmost states, averaging $3,500 per rancher (Callenbach 2000).

Because of these land use changes involving fewer and larger farms and ranches, the High Plains were slowly becoming depopulated during the second half of the twentieth century. Human densities across much of the entire region dropped to six or even fewer people per square mile by 1990. By then it was nearly possible to drive the approximately 1,200 miles from the Mexican border of New Mexico or western Texas to the Canadian border of eastern Montana or western North Dakota without crossing any counties having a human density of more than six people per square mile, although finding food, gas, and lodging during such a long trip might pose a problem. By 1990, 133 counties in the western states had even dropped to a population density of less than two people per square mile, representing true frontier levels (Callenbach 2000). As of 2000 cattle vastly outnumbered people in South Dakota (by roughly a five-to-one ratio) and Nebraska (about four to one), and by smaller proportions in Kansas, North Dakota, Oklahoma, and Wyoming.

This, finally, is the punch line of our two hundred years on the Great Plains: we trap out the beaver, subtract the Mandan, infect the Blackfeet, and the Hidatsa and the Assiniboin, overdose the Arikara; call the land a desert and hurry across it to get to California and Oregon; suck up the buffalo, bones and all; kill off nations of elk and wolves and cranes and prairie chickens and prairie dogs . . .
– *Ian Frazier,* Great Plains *(1989)*

Table 11. Mixed-grass and shortgrass prairie acreage estimates in the Great Plains

STATE/PROVINCE	HISTORIC ACREAGE	PRESENT ACREAGE	DECLINE %
Alberta			
Mixed-grass [a]	21,481,500	8,395,000	67.0
Shortgrass	NA	NA	NA
Saskatchewan			
Mixed-grass [a]	33,086,400	6,172,850	81.3
Shortgrass [b]	14,570,000	2,074,800	85.8
Manitoba [a]			
Mixed-grass	1,481,500	750	99.9
Montana [c]			
Grass/shrub	53,337,496	37,496,108	29.7
North Dakota			
Mixed-grass [d]	35,088,200	11,119,500	68.3
Shortgrass [e]	17,400,000	NA	NA
South Dakota [d]			
Mixed-grass	3,953,600	1,186,080	70.0
Shortgrass	442,309	287,501	35.0
Wyoming			
Shortgrass [d]	7,413,309	5,930,400	20.0
Grass/shrub [c]	28,929,464	25,452,430	12.1
Nebraska			
Mixed-grass [d]	19,026,700	4,694,900	75.3
Sandhills [e]	12,000,000	12,000,000	0
Shortgrass [e]	5,000,000	NA	NA
Colorado [c]			
Grass/shrub	27,223,110	15,621,810	42.6
Kansas [e]			
Mixed-grass	About 24,500,000	NA	NA
Shortgrass	About 9,250,000	NA	NA
Oklahoma			
Mixed-grass [d]	6,177,500	NA	NA
Shortgrass [b]	3,211,000	NA	NA
New Mexico			
Shortgrass [d]	NA	3,101,599	NA
Grass/shrub [c]	44,749,995	42,213,264	5.7

Table 11. continued

STATE/PROVINCE	HISTORIC ACREAGE	PRESENT ACREAGE	DECLINE %
Texas [d]			
Mixed-grass	17,407,400	1,777,800	90
Shortgrass	19,273,800	3,953,600	79.5
Contiguous states [f]			
Mixed-grass [g]	139,844,900	NA	NA
Shortgrass [h]	151,959,300	NA	NA

Note: NA = not available.

[a] Estimates of Samson 1998 (p. 439). Mixed-grass estimates are of fescue grasslands.

[b] Estimates from "The Living Grasslands" Web site: http//biology.usgov/s+t/SNT/noframe/grt139htm.

[c] Includes shrubsteppe acreage (after Luce 2003). Historic acreage estimates are totals of current grass/shrub coverage plus croplands within Great Plains prairie dog range. New Mexico estimates include arid grasslands outside the limits of the Great Plains.

[d] Estimates of Samson and Knopf 1994.

[e] Estimates of author (various sources, including map estimations).

[f] Estimates of Risser et al. 1981, based on map planimetry of A. W. Küchler's (1966) vegetation map.

[g] Overall mean of remaining mixed grass (from data available): about 18 percent.

[h] Overall mean of remaining shortgrass (from data available): about 30 percent.

THE RANCHING ERA OF THE 1900S

Eventually, 75 percent of the publicly owned lands in the West was opened to grazing for next-to-nothing grazing fees. Stock growers used their political power to have the Forest Service, established by President Theodore Roosevelt in 1905, placed under the ranch and farm-friendly Department of Agriculture, rather than the Department of Interior, where it logically belonged. Thus the increasingly large areas set aside as national forest lands were opened to ranchers for their personal exploitation, and ranchers were able to buy grazing rights to federal lands at rock-bottom prices (Stegner 1992). The national forests became increasingly exploited by timber interests during the 1900s, especially in the great forest areas of the Pacific Northwest, where "multiple use" of forest resources (basically logging and mineral exploration) became the watchword of the day.

During the years of the Great Depression, ranchers survived better than the dryland farmers. The number of cattle on the Great Plains hit a low point in 1937 but then began to rise, from an average of 124.2 head per ranch in 1937 to 134.2 in 1939, a 7.9 percent increase, and the average family ranch correspondingly increased in size by 17.9 percent, from 2,890 acres to 3,408 acres. With the start of World War II the fortunes of cattle ranchers greatly improved, with the military buying vast quantities of beef. Beef prices gradually rose throughout the war period, despite meat rationing and price controls. The number of farms on the Great Plains decreased 8.1 percent during the war years between 1940 and 1945, but the average ranch size increased by 3 percent. During the first half of the twentieth century the Great Plains accounted for slightly over a third of the nation's rangelands, and they supported from half to perhaps as much as two-thirds of the range livestock. Ranch sizes on the northern plains continued to increase to an average of 3,990 acres by 1952, with statewide percentage increases of more than 20 percent between 1940 and 1950 in Colorado, Wyoming, Montana, and New Mexico. The number of Great Plains farm and ranch units continued to decease correspondingly, especially in states such as New Mexico, where the decline in units was 30.8 percent between 1940 and 1950 (Schlebecker 1963). Between 1980 and 2000 the three states with the largest percentage of counties having human population declines of more than 10 percent were North Dakota (89 percent), South Dakota (64 percent), and Nebraska (60 percent). Slightly farther down the list were Montana (50 percent), Kansas (50 percent), and Wyoming (48 percent) (data from the Economic Research Service, U.S. Department of Agriculture).

As late as early 2000 the U.S. Bureau of Land Management (BLM) still charged minimal grazing fees in its vast federal land holdings (see Table 12 in Chapter 10) of only $1.43 per animal-unit-month (an AUM typically consisting of an adult cow/calf pair, one bull, one horse, or five sheep being allowed to graze on public land for a month) – a price that was about 20 percent that typical of fees on privately owned lands during the same period. These fees regularly produce less income for the BLM than the amount of money required by the agency to administer its lands (Stegner 1992). No wonder virtually all of the federal lands in the West, except for national parklands and nature preserves where grazing is not permitted or at least is strictly controlled, are severely if not fatally overgrazed.

The average range steer consumes about 6 tons of vegetation before it is ready for the market. Additionally, some 85 percent of our agricultural lands are used for producing animal foods, and an estimated 70 percent of the water in eleven western states is now being used for raising animals to be used as human food. The U.S. government pays an average of $54 per acre to subsidize irrigation on these arid lands. This of course excludes the additional government subsidies paid to cattle ranchers to help hold livestock prices above the prices available from foreign markets and to extricate the ranchers from bad times during drought or other hard times. Public-land ranching costs taxpayers about $1 billion annually yet produces only 3 percent of American beef (Lyman and Metzer 1998).

First, I would get rid of the stinking, filthy cattle. Every single animal. Shoot them all, and stock the place with real animals, real game, real protein: elk, buffalo, pronghorn antelope, bighorn sheep, moose. – *Edward Abbey,* One Life at a Time, Please *(1988)*

THE GROWTH OF BISON RANCHING

The development of commercial bison ranching in America has been extremely slow, in large part because of the numerous political roadblocks that the powerful western cattle interests have repeatedly tried to place in its way. Thus grazing permits on BLM and Forest Service lands are regularly given to cattle and sheep, both of which are far more destructive than are bison to native vegetation, but such permits are denied to bison. Similarly, the low-fee grazing rights for rancher-owned cattle on national wildlife refuges have not been extended to bison.

As mentioned in Chapter 2 the bison (Figs. 34 and 35) was saved from near-extinction around the end of the nineteenth century through the efforts of a few high-profile conservationists, including Theodore Roosevelt and William Hornaday. The American Bison Society was formed in 1905, with Hornaday as its president and President Roosevelt as its honorary president. The primary purpose of the society was to save the species from extinction, in part as a symbol of the vanishing western frontier and also for its potential as a potential domesticated draft animal and a source of high-quality wool. Through the society's efforts several herds were established on federally owned lands, the first in 1907 on 8,000 acres of prairie in western Oklahoma (later to be enlarged and developed as the Wichita Mountains Wildlife Refuge), and the second in 1909 near Missoula, Montana (the National Bison Range). By 1919 new herds had also been started at the Niobrara Game Preserve (now Fort Niobrara National Wildlife Refuge) in Nebraska and at Wind Cave National Park and Custer State Park in South Dakota. The tiny remnant herd at Yellowstone National Park was also being protected effectively and was thriving with supplements from captive herds (McHugh 1972).

The bison industry received its start from the relatively few animals still surviving on private ranches in the early 1900s, through the efforts of a handful of people (mentioned in Chapter 2), including Samuel Walking Coyote, James "Scotty" Philip, "Buffalo" Jones, Michel Pablo, and Charles Allard. An early survey indicated that about seven hundred bison existed in captivity in 1902,

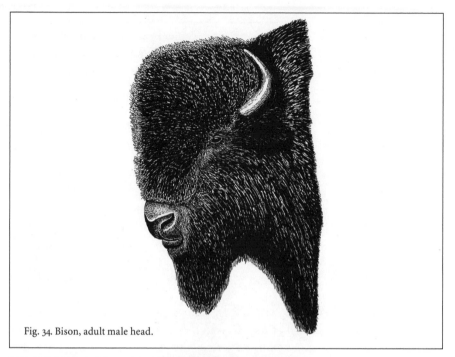

Fig. 34. Bison, adult male head.

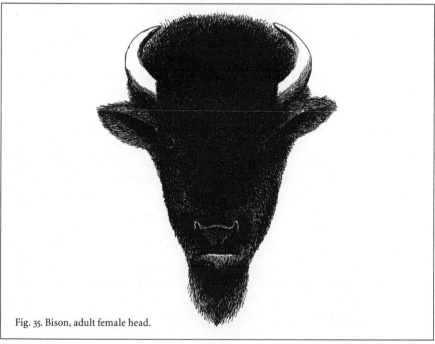

Fig. 35. Bison, adult female head.

a number that slowly increased as the commercial value of bison meat, bison hides, and bison as a tourism attraction became more apparent.

In 1967 the National Buffalo Association was founded, and in 1992 the InterTribal Bison Cooperative was formed among a group of forty-six Native American tribes representing seventeen states (Conley and Albrecht 2000). By the late 1900s more than 225,000 bison were being maintained on private ranches and over 5,000 were owned by tribal groups, roughly fifteen times more than the number then present on state and federal lands.

Until the early 1990s the supply of bison meat for the commercial market was fairly inconsistent and consisted largely of spare bulls and old cows, so meat quality was generally unimpressive. During the 1970s and 1980s the average prices of bison sold at Custer State Park varied from less than $450 to about $1,000, with the highest prices occurring in the late 1980s. During the decade from 1989 to 1998 per-carcass prices increased gradually, with adult animals reaching an average peak price of about $2,750 in 1998. By the mid-1990s carcass prices had jumped from about $1.75 per pound to $2.35 per pound, or well above prices being brought by cattle. With the increased abundance of bison on the market in recent years, meat quality has improved and vaccination against brucellosis has produced a disease-free captive bison herd.

By the end of 1999 there were about 300,000 bison living in the United States and Canada, with about 141,000 in U.S. private herds, 100,00 in Canadian private herds, 10,000 in U.S. public herds, and 3,000 in Canadian public herds. Private herds could be found in all fifty states and all Canadian provinces, as well as in numerous foreign countries. The largest statewide components of the total U.S. bison population were in South Dakota (18.4 percent), Montana (8.7 percent), North Dakota (8.7 percent), New Mexico (8.4 percent), Colorado (7 percent), and Wyoming (5.1 percent), all of which are states in the High Plains ecoregion.

In 1966 bison ranchers established the National Buffalo Association. A separate bison ranching organization, the American Bison Association, was formed in 1975. In 1995 the two groups merged to form the National Bison Association. Similar national organizations exist in Britain, Canada, Finland, and Japan, as well as state or regional organizations in more than a dozen locations. The U.S. Department of Agriculture has agreed to add bison to its annual Census of Agriculture as well as to begin tracking the number of animals processed under USDA and state inspection procedures.

Bison leather is now being used for making very high quality footwear, having a wearable lifetime up to 80 percent greater than ordinary cowhide. The underfur of bison, which is extremely soft and luxurious, is used for making blazers, sweaters, vests, and coats. Almost every part of the animal now finds some commercial use, from decorative skulls to handmade paper formed from dried bison feces.

Raising bison on the arid lands of the West offers a large number of ecological benefits, as outlined in Chapter 2. Bison meat also is far more healthy for human consumption than is beef. For example, even the Department of Agriculture, whom nobody can accuse of not toadying to cattle ranchers, has admitted that the average fat content of bison meat is 1.8 percent, as compared with 6.2 percent for beef and 3.1 percent for chicken. Nor is bison meat laced with the profusion of antibiotics and other prescription medicines regularly given to cattle, of which an estimated 40 to 85 percent have not been approved by the U.S. Food and Drug Administration for the diseases they are intended to treat (Callenbach 2000). Additionally, the U.S. Department of Agriculture has determined that bison meat has lower levels of cholesterol than beef (95 percent relative to beef), pork (95 percent), or chicken (92 percent), and fewer calories per unit weight than any of these (68 percent relative to beef or pork; 75 percent for chicken) (Anonymous 2000).

Hell, this could be a business. – *Ted Turner, commenting on the value of bison relative to beef carcass prices in the 1990s*

10

The USFS, BLM, and BIA

How the West Was Lost

The surface of the earth does not offer a more sterile sight than some dry-land pastures of America with nothing but sheep trails across their grassland grounds. The free-enterprisers of these ranges, many of them public-owned, want no government interference. They ask only that the government maintain trappers, subsidies on mutton and wool, and tariffs against competitive importations. – *J. Frank Dobie*, The Voice of the Coyote *(1949)*

THE U.S. FOREST SERVICE

The U.S. Forest Service (USFS) operates within the U.S. Department of Agriculture and administers 155 national forests occupying nearly 200 million acres of public lands, as well as 20 national grasslands covering 4 million acres, in total encompassing about 8.5 percent of the U.S. land area. It also manages 19 of the nation's 38 national recreation areas, some of its 8 national scenic trails, and 33 percent of the nation's 630 wilderness area. The wilderness areas, which were established by the Wilderness Act in 1964 from existing public land holdings, encompass 102 million acres. The national forest areas alone represent 25 percent of the nation's publicly owned lands. The ten-state High Plains region extending from Montana and western North Dakota south to New Mexico and western Texas contains 17 national grasslands, collectively encompassing 3.9 million acres. Of the total acreage of public lands in the West, including federal, state, and local holdings, the U.S. Bureau of Land Management (BLM) and the Forest Service collectively control about 75 percent (Lyman and Metzer 1998). There are 9 regional Forest Service offices, and 154 forest, grassland, and ranger district offices.

In 1891 Congress set aside the first of numerous forest reserves (the Yellowstone Timber Land Reserve) from public domain land. This reserve and subsequent reserves were originally under the control of the General Land Office in the Department of Interior. In 1900 the National Cattlemen's Association began to lobby for a federal land-leasing bill that would allow unlimited grazing on forest lands. During this period the Department of Interior required cattle ranchers to obtain permits to graze these lands, which offended the ranchers, who evidently believed they should be allowed to do whatever they wished within the forest reserve lands (Schlebecker 1963).

In 1905 these forest reserves were shifted from the control of the Department of Interior to the Department of Agriculture, partly under the lobbying influence of cattle ranchers hoping for better treatment there, and were designated as national forests. During that year Teddy Roosevelt reorganized the Forest Service, naming Gifford Pinchot its director, and for good measure added five national parks and sixteen national monuments, placing these new lands out of reach of the timber barons. In 1906 the reorganized Forest Service established a grazing fee in forest lands, which varied locally but averaged $0.47 per animal-unit-month (AUM). These fees caused dismay and objections among the cattle ranchers, but the Service's decision to levee such grazing fees was upheld by the Supreme Court in 1911 (Clawson 1971).

In 1907 Congress fixed the grazing fees in Forest Service lands at amounts ranging from 25 cents to 35 cents per AUM, or the price of grazing one adult head of cattle (or five sheep) for a month on Forest lands (Schlebecker 1963). The ecological cost of an AUM may be measured as the amount of forage required to feed an animal unit for one month, an amount roughly equal to 800 pounds of air-dry forage. In the shortgrass prairies of Montana, for example, the AUM has been estimated by the BLM as representing 6.1 acres – meaning that 6.1 acres of land are needed there to support one head of cattle. Overall average AUMs for all BLM lands in 1967 were 13.3 acres (Clawson 1971).

Congress passed the Forest Homestead Act in 1906, making homesteading possible in some national forests but further restricting unlimited forest grazing by cattle ranchers. Also in 1906 Congress established a Bureau of Dry Land Agriculture. New homesteading laws passed in 1909 allowed for the settling of 320 acres of dry land and stimulated a new burst of homesteading. During the decade 1900–1910 the number of farms on the Great Plains increased 37 percent, to 1.18 million, and the amount of land under cultivation increased to 324.6 million acres, representing an average of 274 acres per farm. The increasing influence of farming and the impact of the Forest Service on the cattle industry were only just beginning to show their effects (Schlebecker 1963).

Probably the greatest single influence on the development of grazing in the Great Plains came about during the drought years and Great Depression in 1934 with the passage of the Taylor Grazing Act. This act permitted the Department of Interior to administer the remaining 143 million acres of American public domain, allowing for the establishment of grazing districts on some 80 million acres, an area that was increased to the entire 143 million acres in 1936. In that same year the Forest Service estimated that the public domain lands were 67 percent depleted from their virgin condition. A director of grazing was appointed by President Roosevelt, with the tiny grazing fees (less than 5 cents per animal-unit-month, or about 50 cents per head of cattle per year) to be split between the federal government and the individual states where the grazing occurred. These provision provided cattle ranchers with almost all they could possibly ask for, as the grazing fees were less than half the measly 13 cents per AUM then being charged by the Forest Service. By 1937 there were 1.8 million cattle and 7.8 million sheep grazing on 110 million acres of leased public domain lands, with the remaining public lands probably being grazed without benefit of lease. The resulting range degradation and erosion did not worry the stock growers; the secretary of the American National Livestock Association noted reassuringly in 1938 that "erosion there has always been; erosion there will always be" (Schlebecker 1963).

It was apparent by 1940 that the tiny fees charged by the Division of Grazing were not even paying for the costs of administering the operation. By 1945 the Forest Service fees for grazing had risen to 25 cents per AUM, as compared with only 0.05 cent still being charged by the Division of Grazing (Schlebecker 1963). In 1946 the Grazing Service was consolidated with the General Land Office into the Bureau of Land Management.

By the late 1940s the Forest Service reported that its forests and grassland domain consisted of 230 million acres, which were then being grazed by 10 million cattle, and the forest grazing rights were leased by twenty-five thousand ranchers (U.S. Department of Agriculture 1949). That works out to one cow per 23 acres, and an average of four hundred cattle per rancher. In 1987 the Forest Service received $8.1 million in income from its grazing operations but spent $31.4 million. Collectively, it and the BLM had grazing-expense deficits of $50.8 million in 1987 (Callenbach 2000).

The nation's national grasslands are also under the control and management of the Forest Service. They had their origins during Roosevelt's New Deal period of the Great Depression. In 1937 the ill-fated Resettlement Administration of 1935, designed to purchase submarginal lands of the Great Plains and use the money to create new towns, was replaced with the Farm Security Administration. This program similarly allowed for the purchase of submarginal lands but, rather than creating new towns for the displaced persons, was designed to develop a program of land conservation. Some 11.3 million acres of land were thus purchased and reseeded, and appropriate grazing management practices were started. These newly acquired lands were transferred to the Forest Service in 1953. Much of this acreage was later in turn transferred to various state and federal agencies; in 1960 the remaining 3.8 million acres were designated as

Table 12. BLM, national grassland, and homestead acreages in the Great Plains

STATE	BLM LANDS	NATIONAL GRASSLANDS	HOMESTEADS [a]
Montana	8,124,320	None	32,050,480
North Dakota	<50,000	1,104,269	17,417,466
South Dakota	300,000	863,767	ca. 15,660,000
Wyoming	17,543,148	572,319	18,225,327
Nebraska	<50,000	94,332	22,253,314
Colorado	8,331,857	612,000	22,146,400
Kansas	<50,000	108,175	13,089,258
Oklahoma	<50,000	ca. 16,000	14,865,912
New Mexico	12,948,416	136,505	19,422,958
Texas	None	ca. 89,000	None

[a] Homestead acreages as of 1961, after which homesteading occurred only in Alaska (data from Clawson 1971).

twenty national grasslands, mostly located in the Great Plains (Table 12). Additional components of the lands thus obtained were converted into or added onto national wildlife refuges, including additions to Valentine, Crescent Lake, and Fort Niobrara National Wildlife Refuges in Nebraska.

The national grasslands could potentially support major populations of typical High Plains biota, especially such classic shortgrass prairie species as the black-tailed prairie dog, swift fox, burrowing owl, lesser prairie-chicken, and ferruginous hawk. Yet, between 1985 and 1998, the Forest Service poisoned some 97,000 acres of prairie dog colonies in the twelve national grasslands of the Great Plains (Predator Conservation Alliance 2003). These federally owned lands (that is, lands owned by the American public) actually support fewer prairie dogs and associated native fauna than do many adjoining grassland regions, and they largely consist of biologically degraded "national grazelands" that help underwrite large cattle-ranching operations.

The U.S. environmental costs to the natural environment became increasingly evident by the 1960s in the form of polluted water, land, and air, and overexploitation of the land and wildlife resources. The National Environmental Policy Act of 1969 (amplified in 1973), the Endangered Species Act of 1973, and the National Forest Management Act of 1976 began to put some brakes on the systematic destruction of the national forests by the Forest Service. Since 1986 the Forest Service has also been expected to select and attempt to manage for regionally "sensitive" species that exhibit significant current or predicted downward trends in population numbers or density, as well as federally threatened or endangered species. An estimated 30 percent of the nation's federally listed threatened and endangered species are known to exist in the nation's national forests and national grasslands. As of the early 1990s the five-state Rocky Mountain Region of the Forest Service (consisting of most or all of Wyoming, Colorado, South Dakota, Nebraska, and Kansas) was known to support 67 species of endangered, threatened, or vulnerable terrestrial wildlife, including 5 amphibians, 6 reptiles, 22 mammals, and 33 birds. The Northern Region (Montana, North Dakota, and parts of Idaho and South Dakota) had 23 species identified as sensitive, and the Southeastern Region (including New Mexico and Arizona) had 237 species identified as sensitive (Finch 1992).

Every Forest Service site is now subject to the maintenance of viable populations of native vertebrates in national forests, as required by the National Forest Management Act. The Endangered Species Act of 1973 had already required that federal agencies not jeopardize the continued existence of any endangered or threatened species or act in a way that results in the destruction or adverse modification of its habitat. Thus the endangered spotted owl of the Pacific Northwest at least temporarily halted the cutting of old-growth national forests in the Pacific Northwest.

However, ingrained forest-cutting and overgrazing habits of the Forest Service are hard to change. The Forest Service has become the world's largest road-building agency, having built by 1990 some 340,000 miles of logging roads, or eight times as many miles as the collective 42,000-mile length of the U.S. interstate highway system. Building logging roads fragments the forest community, causes land erosion, provides avenues for invasion by alien weeds and animals, and opens previously pristine forest areas to logging and other disturbance. As recently as the early 1990s a timber company in Alaska could still buy a publicly owned tree that might be several hundred years old for less than the price of a hamburger, and in Idaho lodgepole pines were being sold for less than a dollar a tree (Manning 1991).

The Forest Service has admitted losing money on its timber sales in 91 of the nation's 156 national forests, since the costs of building roads and overseeing the logging are greater than the amount of money the timber companies pay the government. The "profits" in the other forests are achieved by cooking the books in various ways, such as amortizing the costs of building roads in the Chugach National Forest of Alaska over a period of 1,200 years (Manning 1991). Although the Forest Service claims to close roads after logging has been terminated, a study in Oregon revealed that only 20 percent of such roads were actually closed (Callenbach 2000).

Like other federal, state, and private organizations, the Forest Service has been increasingly forced to come clean as to the environmental impacts of its actions since the 1973 National Environmental Policy Act. However, the Forest Service has managed to think up some innovative, prospective "categorical exclusions" that would allow it to act without following these rules. One proposed exclusion would permit activities that allow forest thinning, prescribed fires, and livestock grazing regardless of the size of the acreage affected. Another proposed exclusion would exempt a citizen's right to appeal all decisions that implement forest plans for so-called small timber sales. The Forest Service has also proposed limiting overall citizen oversight rights as to the kinds of decisions that might be appealed. Additionally, the changes proposed would exempt from appeal any change made directly by the secretary or under secretary of agriculture. Other changes further freeing the Forest Service to allow additional uncontrolled forest cutting are being proposed by the Bush administration (Stahl 2003).

The Forest Service estimated in the late 1990s that, as of 2000, its annual income from logging would be $3.5 billion, as compared with $110.7 billion from recreation users. Yet lumbering continues to rank much higher in its priorities than conservation and forest maintenance activities that attract nonconsumptive users (Brower 1997).

THE BUREAU OF LAND MANAGEMENT

The Bureau of Land Management was established in 1946 as a direct descendant of the Taylor Grazing Act of 1934 and the associated federal Grazing Service. The BLM operates within the Department of Interior. Its principles of multiple use include livestock grazing, timber, industrial development, watershed protection, mineral production, outdoor recreation, and fish and wildlife production. There are 12 BLM state offices, 59 district offices, and 170 resource area offices. Each district has an advisory board of up to thirteen members, all but one of whom are stock growers. The state advisory boards were once composed of nothing but stock growers but were expanded in 1961 to represent additional opinions. The national advisory board likewise historically had only livestock representatives, but it too was expanded in coverage in 1961 to include the consideration of wildlife interests and other alternative viewpoints (Clawson 1971).

The BLM administers 270 million acres in the western states, making it the largest single manager of all federal lands. The BLM also oversees the mineral rights to some 570 million acres. Five percent of the nation's 630 wilderness areas are also under the jurisdiction of the BLM. Of the total land area administered, 173 million acres consist of designated grazing lands, including 134 million acres of publicly owned grazing-district lands located in ten western states, 20 million acres of reserved lands or lands within grazing districts that are owned by public organizations, and an additional 19 million acres of grazing lands outside these grazing districts that are under lease to individual grazers. These total grazing lands represent an area about twice the size of California. In the five-state region of the High Plains from Montana south to New Mexico, the BLM controls more than 49 million acres, representing approximately the same area as the entire state of South Dakota.

Although not so significant in the High Plains region, the BLM also has vast forest holdings in the western states, much of which were cut illegally while they were still in the public domain and before they could be protected effectively from logging. In states other than Alaska and Oregon, the BLM administers some 2 million acres of forest, representing about 9.5 billion board feet of lumber. All told, the BLM manages about 134 million acres of forests, representing some 271 billion board feet of lumber. The BLM also sells a vast amount of timber taken from land under its control, amounting to about a billion board feet annually as of about 1970 (Clawson 1971).

The Bureau of Land Management had its nineteenth-century origins in the federal government's General Land Office, which was given the mandate of distributing homestead lands, the leftover lands that were not taken over as forest reserves or otherwise eventually exploited for free and virtually unlimited grazing by sheep and cattle, as described above. An intermediate stage of governmental control was the federal Grazing Service, established in 1934. This in turn was succeeded by the present-day Bureau of Land Management in 1946. In its first full year of operation the BLM collected $21 million (and spent about $5 million), largely through mineral development, forestry, and grazing fees. By 1965 the Bureau took in $244 million and spent $21 million annually (Clawson 1971).

In 1964 the Bureau's authority was expanded by the passage of the Classification and Multiple Use Act, which allowed the BLM to classify public lands into three categories. Type I lands are those with large acreages likely to be managed indefinitely. Type II lands are those in smaller blocks, scattered among private lands, which may be either retained or disposed of. Type III lands are small or scattered tracts that are often considered disposable. This 1964 act and other more recent ones have required the BLM to protect wilderness areas while at the same time making their designated recreational areas more generally available.

In 1967 an estimated 49 million visits were paid to BLM land for recreational purposes, at a time when only 160 developed camping and picnic sites existed. In 1968 there were about 15,000 permits issued for grazing livestock in grazing districts, with 80 percent for cattle and horses and the remainder for sheep and goats. Another 8,600 permits were granted for grazing areas outside designated grazing districts, generally using unfenced lands adjacent to the lessee's privately owned lands. The average permittee within the grazing districts had 725 AUMs of grazing use, and those outside the districts averaged 310 AUMs of grazing use (Clawson 1971).

In 1973 Congress passed the National Environmental Policy Act, requiring federal agencies to truthfully acknowledge the environmental impacts of their activities. In 1975 about 85 percent of the BLM land was classified by that agency as being of "unsatisfactory" quality, an abysmal figure that the General Accounting Office considered to be an underestimate (Lyman and Metzer 1998). The BLM's powers were further defined and limited by the Federal Land Policy and Management Act of 1976, when the agency increasingly was required to listen to the public's concerns for land use practices other than maximizing grazing and timber cutting on BLM lands. By 1984 the BLM reported to Congress that 60 percent of the land under its stewardship was in poor condition (Abbey 1988), a seemingly slight improvement from the conditions reported during the 1970s. However, during the every-capitalist-for-himself years of the 1980s, James Watt, Ronald Reagan's secretary of Interior, turned the leadership of the BLM over to Robert Burford, and the foxes had truly been given the keys to the chicken coop (Stegner 1992).

It did not take Watt long to take action against the "environmental extremists." In 1982 he excluded 1.4 million acres of wilderness study areas from further consideration and instead opened the areas to exploration and development. He also approved coal leases on BLM land for giveaway prices, including leases on 1.1 billion tons of coal in Montana and Wyoming for $43.5 million, or about $100 million under market value. A second enormous sale occurred in western North Dakota and Montana, at a price so low that the House Interior Committee asked a federal court to invalidate the sale. By 1984 Watt had made deep cuts in the BLM budgets for wildlife habitat, wilderness, land use planning, and soil, air, and water management and had begun to pack its advisory boards with people favoring resource exploitation. As a result of a 1983 federal suit against him by the Sierra Club, the wilderness study areas were again restored for consideration and Watt was effectively forced out of office (Zaslowky and the Wilderness Society 1986). In 1986 Dyan Zaslowski summarized the dismal situation in the BLM by saying that "the influence of the Western stockman, the Western miner, the Western developer and the Western politician still obstructs the agency's exercise of what power it does possess."

The Clinton administration attempted to rein in the environmental excesses of the Reagan years, but their efforts, such as an attempt to raise grazing fees on public lands in 1994, were regularly met with defeat by a Republican Congress. The director of the BLM during the early Clinton years, James Baca, tried to change the ways his managers were rewarded from the number of cattle grazed and the miles of fences built to environmental improvements made, but he was soon forced to resign under pressure from western governors (Callenbach 2000). In 1990 only 15 percent of the grazing permits on BLM land were "ancestral," and 3 percent of the permit holders controlled 40 percent of the grazing rights to public lands. Only 10 percent of the grazing rights on all BLM were extended to small operators (Jacobs 1991).

The BLM has made no real effort to document the prairie dog populations on its holdings, except in Montana. There, in the Phillips Resource Area of Phillips County, where black-footed ferret reintroductions are planned or under way, prairie dog colony acreages approached 1 percent of the landscape area by the early 2000s.

I also suggest that we open a hunting season on range cattle. I realize that beef cattle will not make sporting prey at first. Like all domesticated animals (including most humans), beef cattle are slow, stupid and awkward. But the breed will improve if hunted regularly. And, as the number of cattle decrease, other and far more useful, beautiful and interesting animals will return to the rangelands and will increase. – *Edward Abbey,* One Life at a Time, Please *(1988)*

THE BUREAU OF INDIAN AFFAIRS

One does not sell the earth upon which people walk. – *Tshunca-Uitco (Crazy Horse)*

The U.S. Bureau of Indian Affairs BIA operates under the Department of Interior and administers 55 million acres of reservation lands, consisting of more than three hundred federal reservations and twenty-one small state reservations. The BIA has twelve regional offices as well as local agencies on many reservations. Within the High Plains ecoregion, there are major reservations in Montana (Fort Peck, Crow, Northern Cheyenne, and Rocky Boy), South Dakota (Cheyenne River, Crow Creek, Lower Brule, Pine Ridge, and Rosebud, plus most of Standing Rock) and North Dakota (Fort Berthold and part of Standing Rock).

In 1824 the Office of Indian Trade (which had been established in 1806) was replaced by an Office of Indian Affairs in the War Department. The agency's name was later changed to the Bureau of Indian Affairs, and the agency transferred to the Department of Interior. Most of the Indian reservations were established after the Civil War, especially during the period of western expansion between 1867 and 1887.

For example, in 1877 the Great Sioux Reservation, consisting of 22.4 million acres in what is now South Dakota, was established, but in 1888 – under pressures from land speculators and railroad interest – it was broken up into six smaller reservations, with the excluded areas providing about 9 million acres for non-Native settlement. The Lakota Sioux had been initially offered 50 cents per acre for this land, an offer that they rejected. The former reservation lands that were opened to homesteading proved to be mostly unsuitable for farming, with only 5 percent of the land "proving up," and over a quarter of it soon reverting back to the government. Reservation breakups and forced tribal displacements by the Pawnee, Ponca, and other Great Plains tribes were common during this period, with the tribal members often being removed permanently to new reservations in Indian Territory (Oklahoma).

The General Allotment (Dawes Severalty) Act of 1887 assigned many of the remaining traditional tribal lands to individual Native Americans, with 91 million acres that were designated as surplus becoming available for purchase or homesteading to non-Natives. However, the Natives often later lost their individual allotments through tax defaults. The Curtis Act of 1898 further reduced Indian reservation lands from about 150 million acres to 60 million acres (Waldman 1985). Still later reductions have diminished the reservations to slightly more than 52 million acres, an area about equal in size to the state of Nebraska. Nine Indian reservations, totaling 12.6 million acres, are present in the region designated in this book as the High Plains.

During the early 1900s Native Americans on the High Plains were ordered by the BIA to begin establishing cattle herds on their reservation lands, but in 1917 tribes such as the Lakota Sioux were instead encouraged to sell their herds and lease the lands to non-Natives. After the lessees defaulted on their leases, the government agents urged the Natives to sell their land allotments, thus losing both their land and their cattle (Pritzker 1998).

However, under the Roosevelt administration, the Indian Reorganization Act of 1934 reversed the federal government's long-held exploitive policies of land allotments and the assimilation of Native Americans into the general population and instead returned unsold allotment lands to tribes, provided legal sanctions for tribal lands, and provided for the purchase of new lands. It also encouraged the hiring of Native Americans by the BIA, restored religious freedom to Native Americans, and advocated the education of children through reservation schools, rather than through off-reservation boarding schools where Native languages, cultures, and tribal histories were not encouraged or taught (Waldman 1985).

The Indian Claims Commission, established in 1946, helped to compensate Native Americans by allowing them to sue the government for past treaty violations. They were awarded some $800 million in compensation for lost lands until the commission was terminated in 1978. Despite these advances the policies toward Native Americans during the Republican-controlled era of the 1950s again hardened, with Congress terminating federal relationships with sixty-one tribes, bands, and communities, often as a means of allowing non-Native interests to gain control of rich timber resources on tribal lands. The Termination Act of 1954 eliminated official U.S. government–tribal relationships, offering adult tribal members the choice of taking an acreage allotment or accepting a cash payout. These acts helped to ensure the process of tribal termination and partially compensated tribes for lost lands with token dollar amounts worth much less in the long run than the land itself (Waldman 1985).

For example, the Pine Ridge Lakotas, living on one of the most poverty stricken reservations in the United States, were offered (in 1980) a payment of $17.5 million (plus nearly a century of interest) in retroactive compensation for the overt federal treaty breaking and stealing in 1868 of 7.7 million acres in western South Dakota, including their sacred Paha Sapa, the Black Hills. The Lakotas scornfully rejected the offer, which with accrued interest had reached more than $200 million by the late 1990s (Pritzker 1998). This recent figure represents about $26 per acre for a region that is unusually rich in timber, gold, and other mineral, land, and water resources.

The Wheeler-Thomson (or Indian Reorganization) Act of the 1930s allowed Natives to set up their own governments and to negotiate with the U.S. government. However, it also undermined the traditional tribal structures, and in the 1950s the federal government again began forcibly relocating Native Americans, mostly from reservations to urban centers that often became ghettos (Pritzker 1998).

During the 1960s the federal government's policy of termination ended, and in 1961 the findings of three commissions (the Keeler Commission on Indian Affairs; the Brophy Commission on Rights, Liberties and Responsibilities of the American Indian; and the United States Commission on Civil Rights) began to address the many failures of the BIA as well as the federal government itself in their shameful treatment of Native Americans. Finally, in 1977, the American Indian Policy Review Commission took a hard stand against forced assimilation and promoted the concepts of tribal self-determination and self-government of tribes (Waldman 1985). The current Indian reservations of the Great Plains are now one of the major remaining natural reserves of High Plains wildlife and native plant life, especially for otherwise generally unprotected and vulnerable species such as prairie dogs.

By the end of the twentieth century the Native American population had reached about 1 million, with a small minority living on reservations, as compared with a population of about 250,000 a century before, when most were still confined to about two hundred reservations. Native American residents on the High Plains reservations of Montana and the Dakotas totaled about 45,000 persons in 1990.

Unfortunately, some reservations still engage in massive prairie dog poisoning efforts to support their ranching economies, although they are also pressured by more traditionally minded members to try to maintain the grassland's original biota. The Bureau of Indian Affairs has thus been responsible for some of the largest and most expensive prairie dog poisoning projects in recent decades, an ironic fact given the desperately poor economic conditions of the people living on their reservations. Between 1980 and 1984, $6.2 million was spent in poisoning 458,618 acres of prairie dogs on the Pine Ridge Indian Reservation, at a cost of about $3 per prairie dog killed, and in 1985–86, about 240,000 acres were re-poisoned. The costs of such poisoning far outweighed the value of the forage thus preserved (Predator Conservation Alliance 2003). Surveys in 2003 by Tim Vosburgh, the black-tailed prairie dog tribal coordinator, indicated that 128,000 acres of occupied prairie dog acreages then existed on Native American reservations, within a total area of 1,558,337 acres of current potential habitat. Of the five large prairie colonies (greater than 10,000 acres) remaining in the United States, four primarily occur on tribal lands, indicating that the future of prairie dogs may largely depend on the values that Native Americans place on them.

OTHER APPROACHES: THE NATIONAL PARK SERVICE AND THE U.S. FISH AND WILDLIFE SERVICE

The National Park Service

The National Park Service (NPS) manages 385 federal sites over some 83 million acres, including 57 national parks, 68 national monuments, at least 19 national recreation areas, 13 national preserves, 6 national rivers or wild and scenic riverways, 4 national lakeshores, and a national scenic trail. The Park Service also administers 42 percent of the nation's 630 wilderness areas as well as a large number of historic sites, such as historic national battlefields, national memorials, national historic trails, and national historic parks. The first national park, Yellowstone, was established in 1872, and the first national monument, Devils Tower, was designated by Theodore Roosevelt in 1906. Using the Antiquities Act of 1906 as his personal tool, Roosevelt also established Petrified Forest National Monument and Grand Canyon National Monument, both later upgraded to national park status. The NPS was not formally organized until 1916.

National parks lying within the High Plains region include Theodore Roosevelt National Park in North Dakota, and Wind Cave and Badlands national parks in South Dakota. These High Plains sites collectively comprise about 400,000 acres and include important reserves for bison and black-tailed prairie dogs. Both species are well represented in the first three national parks just mentioned, but both have been extirpated from Carlsbad Caverns National Park, in New Mexico.

Following up on Theodore Roosevelt's innovative use of the Antiquities Act to single-handedly generate new national monuments without trying to obtain congressional approval, President Jimmy Carter established the Bering Land Bridge, Cape Krusenstern, Misty Fiords, and Aniakchak national monuments in Alaska, and President Bill Clinton similarly established Grand Staircase–Escalante National Monument in Utah. Several national monuments are found in the High Plains region, including Little Bighorn Battlefield in Montana; Agate Fossil Beds, Scotts Bluff, and Chimney Rock in Nebraska; Capulin Mountain and Fort Union in New Mexico; Jewel Cave in South Dakota; Alibates Flint Quarries in Texas; and Devils Tower in Wyoming. Their collective areas total about 10,000 acres. National monuments in the High Plains that support or have recently supported prairie dog colonies include Devils Tower, Scotts Bluff, Chimney Rock and, possibly, Agate Fossil Beds. None of these monuments currently supports bison herds.

Hunting is not allowed in national monuments or national parks. At least in theory, lethal control of predators or other "undesirable" wildlife is also not permitted. However, the American white pelicans breeding on Yellowstone Lake were once believed to compete with recreational trout fishermen there, and their breeding colony on the lake was regularly raided by park personnel to destroy the eggs. And, as part of their "good neighbor" policy toward nearby ranching interests, poisoning of prairie dogs has frequently occurred within national parks and national monuments. Between 1985 and 1998 the National Park Service poisoned at least 5,500 acres of prairie dog colonies on national park properties within the Great Plains (Predator Conservation Alliance 2003). In contrast, the reintroduction of wolves into the Yellowstone-Teton ecosystem is an example of a forward-thinking ecological approach to the problem of no longer having any natural predators to control the increasingly large herds of large grazing animals in such sanctuaries.

Historically, twelve national parks supported prairie dog populations, but by the year 2000 that number had been reduced by poisoning or other influences to seven, with a collective colony acreage of about 6,600 acres. The most important of these are Badlands National Park, with about 4,300 colony acres in 2000, Wind Cave National Park, with about 1,600 acres, and Theodore Roosevelt National Park, with about 847 acres (Predator Conservation Alliance 2003).

Other High Plains sites administered by the National Park Service include

Mount Rushmore National Memorial in South Dakota and the Lake Meredith National Recreation Area in Texas. The Pine Ridge National Recreation Area in Nebraska is administered by the National Forest Service. These areas collectively occupy about 100,000 acres. Hunting is allowed on this and many other western national recreation areas, which are variously administered by the National Forest Service, the National Park Service, or the BLM. Prairie dog colonies are present in both the Pine Ridge and Lake Meredith areas, and – at least in the Pine Ridge area – they receive no protection from recreational hunting. These two areas could also easily support managed bison herds rather than allowing continued grazing by cattle.

The U.S. Fish and Wildlife Service

The U.S. Fish and Wildlife Service USFWS is operated under the Department of Interior and administers over 100 million acres of government lands as national wildlife refuges, national fish hatcheries, and wetland management districts. The USFWS also administers 20 percent of the nation's 630 wilderness areas. The first national bird sanctuary, Pelican Island Bird Reservation (now one of the national wildlife refuges), was established off the east coast of Florida in 1903 by executive order during the Theodore Roosevelt administration. Roosevelt followed this precedent-establishing action by naming fifty-four additional bird reservations and game preserves, including the National Bison Range in Montana. The USFWS was formed in 1940 by merging the Bureau of Biological Survey and the Bureau of Fisheries, both previously part of the Department of Agriculture, into a new agency within the Department of Interior.

The USFWS's refuge lands are now scattered over nearly every state, including (as of 2003) 540 national wildlife refuges, numerous wetland management districts in the northern plains, and several national fish hatcheries. The lands also include many federally controlled waterfowl production areas and a few sites that are still listed as national game preserves. One such game preserve is Sullys Hill in North Dakota, an area set aside by President Theodore Roosevelt in 1904 for preserving big game; it was officially named a national game preserve in 1917.

Overall, the federal refuge system contains about 95 million acres, an area almost equal to the entire state of California in size, with 80 percent of the refuge lands in Alaska. There are seven regional USFWS offices and hundreds of local refuge offices, fishery resources offices, ecological services offices, and other administrative offices. Most of the High Plains states fall within the Mountain-Prairie Region (Region 6), including Colorado, Kansas, Montana, North and South Dakota, Nebraska, and Wyoming, but New Mexico, Oklahoma, and Texas are administered within the Southwest Region (Region 2).

Collectively, over 4.5 million acres are represented in more than thirty national wildlife refuges within the mapped ten-state High Plains region. The High Plains region includes some of the largest national wildlife refuges in the United States south of Alaska, such as Charles M. Russell National Wildlife Refuge in Montana and Wichita Mountains Wildlife Refuge in Oklahoma. North Dakota has the most national wildlife refuges of any U.S. state, many of them important prairie preserves for mixed-grass wildlife and plant species.

In a few of the Great Plains refuges, managed bison herds have been established and have either supplemented or variably replaced cattle grazing. Wichita Mountains Wildlife Refuge and Fort Niobrara National Wildlife Refuge were both established (in 1908 and 1913, respectively), at least in part, to serve as bison preserves while the species was still in danger of extinction and later were expanded in area and function as general wildlife refuges. Prairie dogs are present on many of the national wildlife refuges of the High Plains. Hunting of prairie dogs is allowed on some of them; 300 of the nation's total 540 refuges allow for recreational hunting of some kind, and fishing is permitted on 260 of them.

Although the Lacey Act of 1900 helped to set the initial stage for the federal regulation of bird harvests in the United States, American birds received their first adequate federal protection under the International Migratory Bird Treaty between Great Britain (on behalf of Canada) and the United States. The treaty was signed in 1916 and became U.S. law in 1918. In 1936 the same degree of international migratory bird protection was extended south to include Mexico. Enforcement of the provisions of these treaties and laws in the United States is primarily the responsibility of the USFWS, especially its Office of Migratory Bird Management.

A separate act provided for the creation of a network of migratory bird refuges, with 138 refuges being established in the two decades between 1921 and 1940. As noted earlier the U.S. Fish and Wildlife Service was formally organized in 1940. Many enormous Alaskan refuges were also established following Alaska's statehood in 1959, such as the Arctic, Yukon Delta, and Togiak national wildlife refuges. These refuges preserve some of the best remaining wilderness lands and wildlife habitat left in America, and their designation as refuges has (as of 2003) prevented or at least forestalled their exploitation for oil and mineral extractions.

A major improvement in refuge management occurred with the passage of the 1997 Refuge Improvement Act, which stated that priority uses for the refuges are to be wildlife related, namely hunting, fishing, wildlife observation, and nature photography as well as environmental education. Hunting, which was first allowed on national wildlife refuges in 1924, has gradually expanded through most of the refuge system. Nonwildlife activities that previously had commonly been allowed on refuges – including grazing, farming, water-skiing, and off-road vehicle use – are now determined on a refuge-by-refuge basis. In 1997 a consortium of nonprofit agencies and conservation groups formed the Cooperative Alliance for Refuge Enhancement (CARE), which is raising money to address a long list of deferred refuge maintenance needs with an estimated cost of $2 billion.

The passage of the Endangered Species Preservation Act (ESPA) of 1966, the Endangered Species Conservation Act (ESCA) of 1969, and the Endangered Species Act (ESA) of 1973, with amendments in 1978, 1982, and 1988, were landmarks in the American conservation movement. The 1966 ESPA allowed in part for the acquisition of lands to protect endangered species; the 1969 ESCA extended protection to species endangered worldwide, prohibiting their import or sale in the United States; and the 1973 ESA required all federal agencies to undertake programs leading to the conservation of federally listed species. The various amendments that followed further refined and defined the earlier acts' provisions. The 1973 Canada Wildlife Act has provided similar measures of protection for threatened and endangered wildlife in Canada.

Since the passage of these acts, the role of the U.S. Fish and Wildlife Service has become much more visible nationally, and its conservation activities that impinge on private agricultural and timbering interests have become far more controversial. Probably its greatest successes have come in helping to save from possible extinction such widely recognized flagship conservation species as the whooping crane, peregrine falcon, and black-footed ferret. However, the back-log of "candidate" species – those species whose populations and population trends warrant federal listing but for which adequate funds are not available to list them and begin active restoration programs – has long greatly outnum-bered those species that are actually listed. Simply getting a species listed as endangered has been no guarantee of saving it; the ivory-billed woodpecker and Eskimo curlew provide sad examples. As of 2003 there were nearly 400 listed species of endangered U.S. animals and nearly 150 threatened species.

The Office of Endangered Species (OES) holds the legal responsibilities associ-ated with these congressional acts, evaluating and classifying species as potential candidates for listing as threatened or endangered. Until recently, candidate species had been classified as Category 1 (those species for which substantial in-formation exists to support their listing as threatened or endangered), Category 2 (those that may eventually also be listed, pending further investigation), or Category 3 (those that have been removed from consideration for various rea-sons, including having become extinct). Recent candidate species include such typical Great Plains wildlife as the swift fox, the mountain plover, the migrant race of the loggerhead shrike, the lesser prairie-chicken, and the black-tailed prairie dog. Other Great Plains species that are nationally declining and may eventually also deserve federal listing include the burrowing owl, ferruginous hawk, northern harrier, upland sandpiper, and long-billed curlew, as well as most of the shortgrass and mixed-grass passerine birds (Johnsgard 2002).

In August 2004 the U.S. Fish and Wildlife Service deleted the black-tailed prairie dog from its list of candidates for national listing as threatened species. The service made its decision on the basis of recent surveys involving ten states and several Native American tribes, suggesting that active prairie dog colonies now cover 1.8 million acres, or three times the agency's 2000 estimate. However, state agency surveys often overestimate prairie dog populations. Aerial surveys in Nebraska during 2003 by Nebraska Game and Parks Commission personnel showed an estimated 136,991 acres of active prairie dog towns along with another 102,828 acres of "possibly active" colonies (Bischof, Fritz, and Hack 2004). University-sponsored ground surveys taken in 2003 supplemented by aerial photography data from various periods up to 1993 produced a 2003 statewide occupied range estimate of 59,974–86,588 acres (Roehrs 2004). This estimate represents no more than 44–63 percent of the state's contemporary estimate of active colonies, or as little as 25 percent if the colonies described as "possibly active" are also included.

By listing black-tailed prairie dogs, USFWS will not only be taking a legally mandated step to conserve a threatened species, it will also be fulfilling the highest purposes of the ESA by protecting a key component of natural short-grass ecosystems. – *National Wildlife Federation (in its 1998 petition to list the black-tailed prairie dog as threatened or endangered nationally under the provisions of the Endangered Species Act)*

11

The Great Plains Grassland Ecosystem

Can It Be Saved?

In many sobering ways, the histories of the native inhabitants of the American western plains resemble a Grecian tragedy that has been replayed time after time, with only minor variations. When Meriwether Lewis and William Clark ascended the Missouri River and first entered the Great Plains in 1804, the numbers of bison that roamed the plains of the Upper Missouri drainage were incalculable, and the same may be said about elk, pronghorn, and deer. Dozens of tribes of Native Americans were living along the Missouri River, with some villages having thousands of inhabitants.

William Clark tried to provide estimates of the number of persons associated with all the tribes they encountered, as well as for other tribes not encountered, from whatever sources he had available. His total population estimate representing tribes for which he had any direct or indirect knowledge within the general Great Plains region was more than thirty-thousand "soles." The actual total Great Plains population of Native Americans during presettlement times can only be guessed at but may well have exceeded one hundred thousand. By the mid-1800s, after the devastating scourges of smallpox and other introduced diseases had taken their terrible toll but prior to the Indian Wars, the Plains tribal populations may have totaled about 75,000 people (Utley and Washburn 1977).

For most if not all of these Great Plains tribes, the bison represented the central component of their economy and also served as a basic element of their religious symbolism. The number of bison occurring across the Great Plains certainly was in the tens of millions, with various estimates ranging up to 30 million or more. Lewis and Clark's expedition alone killed over two hundred of the animals as a convenient source of food while they were within the species' Great Plains range (Burroughs 1961).

Being small and having little or no human food or pelt value, prairie dogs were ignored by early fur traders. Their numbers in the central and northern Great Plains probably were in the hundreds of millions, if not in the billions. After first encountering black-tailed prairie dogs near the mouth of the Niobrara River, Lewis and Clark passed many large colonies of them, especially in what is now North Dakota and eastern Montana. Like the stones along the river's shoreline, their numbers were considered impossible to estimate.

All three of these inhabitants of the Plains – the Native Americans, bison, and prairie dogs – were soon to suffer the effects of exploration and exploitation under the impact of European settlement. The Native Americans of the Great Plains had already begun to suffer from smallpox and syphilis by the time of Lewis and Clark's expedition, as a result of earlier contacts with Missouri River fur traders. The effects of these diseases soon ravaged the villages of the Upper Missouri basin. Lewis and a few of his men were also responsible for the deaths of one or two Native Americans after some Blackfeet braves tried to steal their horses. This hostile encounter was the first recorded skirmish in what would eventually develop into the Indian Wars of the late 1800s, resulting ultimately in the military conquest of the entire Native American population and the destruction of its cultures. The survivors of this massive genocide were rounded up, forced into confined squalid camps, and eventually concentrated in remote, desolate areas far from their homelands, where they were systematically deprived of their languages, religions, and value systems.

By 1890 the Native American Great Plains reservation population (including Montana and the central Great Plains states from North Dakota to Texas but excluding Oklahoma) was 47,327 people (Prucha 1990). By 1890 Oklahoma Territory had been established, having previously been classified as Oklahoma Indian Territory, with more than 64,000 relocated Native Americans incarcerated there. Except for some Pawnees, Poncas, Otoes, and Missourias, this reservation total was composed of more easterly tribes.

After the Civil War was over and newly constructed railroads offered an easy means of penetrating the western plains, the millions of Great Plains bison were accessible and increasingly decimated by hordes of market hunters and thrill-seeking sportsmen. The last bison herds had been hunted down and essentially eliminated by the 1880s, partly for their meat and skins and also as the military services' means of waging indirect warfare on the Native Americans, whose fortunes largely depended on bison. By 1900 the wild population of bison south of Canada numbered only a few hundred. Like the Native Americans of the Great Plains, the few survivors were mostly confined in captivity, where their opportunity for movement was eliminated and their inherent survival senses and spirit of independence would slowly be bred out of them.

The prairie dog avoided the fates of the bison and the Native Americans for a somewhat longer period. It was not until the early 1900s that full-fledged biological warfare was declared on these rodents by the federal government and poisoning of the Great Plains landscape began in earnest. Unlike the bison and the Native Americans, the reproductive potential of prairie dogs helped reduce the rate of its population decline, and almost a century of intensive efforts by federal, state, and private entities was required to bring the prairie dog to near if not complete extirpation over nearly all of its Great Plains range.

Now, a large proportion of the current residents of the Great Plains have probably never seen a live prairie dog or a bison, except perhaps in a zoo or wildlife refuge. A visiting European tourist whom I recently encountered and who wanted to see actual Native Americans was driven through the Omaha Indian Reservation and its village of Macy, which now cannot be distinguished readily from adjacent nonreservation lands or towns. Not a teepee, bison, or prairie dog was seen there. Like the bison and prairie dog, Native Americans have become virtually invisible in our society and, tragically, have become almost insignificant in modern American cultural awareness.

Following World War II the wholesale destruction of native grasslands accelerated. Under the postwar population explosion and economic expansion, new generations of farming equipment, herbicides, insecticides, pesticides, rodenticides, and fertilizers quickly transformed the Great Plains into a few gigantic agricultural monocultures. The environmental movement of the mid-1900s was ushered in by early ecological warning flags raised by such people as Rachel Carson, with her book *Silent Spring*, and Paul Erlich, with his fateful visions of resource depletion. These warnings were supplemented with eloquent writings by such naturalists as the wildlife biologist Aldo Leopold and the plant ecologist Paul Sears. Powerful conservation efforts were begun by such environmental activists as Mardy and Olaus Murie, who advocated the preservation of wilderness and wilderness-dependent wildlife in the West and Alaska. Also present were the firebrand eco-activists, such as Edward Abbey, and the defenders of generally unpopular predators, including grizzly bears (John and Frank Craighead), wolves (David Mech), and coyotes (Dayton Hyde and Hope Ryden).

At the same time, and just as importantly, land acquisition or conservation easements on native grasslands were begun by a few committed people, including Katherine Ordway, who invested her personal fortune into saving remnant prairies, and by nongovernmental and nonprofit environmental groups, such as the Nature Conservancy and the Audubon Society. Various state and federal agencies, partly under legal pressures resulting from the provisions of the Endangered Species Act of the 1970s, also began trying to preserve the few remaining intact pieces of the Great Plains grasslands and their increasingly rare biota.

As of the early 2000s the natural surviving ecosystems of the Great Plains are few and far between. According to the Northern Plains Conservation Network, or NPCN, less than 1.5 percent of the region consists of national parks, national wildlife refuges, or other areas designated primarily for preserving biodiversity. By comparison, the percentage of land represented nationally by preserved natural habitats in an ecologically enlightened country such as Costa Rica approaches 20 percent. The NPCN has identified the ten largest remaining terrestrial landscapes within the northern portion of the Great Plains region that offer the greatest hope for extensive grassland and grassland biota preservation. Eight of these areas currently support prairie dogs, eight support swift foxes, and two support black-footed ferrets. Mountain plovers are present in two, ferruginous hawks occur in four, and greater sage-grouse occur in seven. The ten selected areas are as follows:

1. Sage Creek, Alberta, and southwestern pastures, Saskatchewan. About 2 million to 2.5 million acres of grasslands along the Alberta-Saskatchewan border.
2. Grasslands National Park, Saskatchewan, and BLM Bitter Creek area, Montana. About 2 million to 2.5 million grassland acres along the Canada-Montana border.
3. Montana Glaciated Plains, north-central Montana. About 2.5 million to 3 million acres of grasslands (and wetlands), northwest of Fort Peck Reservoir.
4. Little Missouri Grasslands and Theodore Roosevelt National Park, North Dakota. Federal grasslands.
5. Terry Badlands, Montana. BLM grasslands along the middle Yellowstone River (Prairie County).
6. Big Open, south-central Montana. Mostly privately owned grasslands between the Musselshell and Yellowstone Rivers.
7. Thunder Basin, Wyoming, and Oglala Grasslands, Nebraska. Federal grasslands in the three-corners area of Wyoming, South Dakota, and Nebraska.
8. Slim Buttes, (Harding County) South Dakota. A mixture of private and public (Custer National Forest) lands in northwestern South Dakota.
9. Badlands National Park, Buffalo Gap National Grasslands, and Conata Basin, South Dakota. Federal grassland lands.
10. Hole in the Wall region, Wyoming. Mostly BLM arid steppelands between the South Fork of the Powder River and the Bighorn Mountains.

It would require a Herculean effort to bring all of these regions into conservation protection, for many are still in private hands or are at least partly controlled by interests other than conservation-minded ones. But a few large ecosystems, rather than many tiny fragments, will be needed if animals requiring large yearly home ranges are to survive in anything like their historic environments and if adequately large gene pools can be maintained to preserve the genetic diversity of rare species. There can be no doubt that the American public appreciates the visual and other aesthetic pleasures of seeing unrestrained wildlife in natural environments, and the value of the genetic potential for future human use represented by the thousands of species of plants and animals thus surviving cannot yet be measured.

The United States has a legacy of land exploitation and habitat destruction of nearly three centuries. This legacy has not yet proved fatal only because of the seemingly unlimited bounty of resources that the country originally possessed. We are now approaching the end of this period of apparent riches, with water, land, and air pollution, climatic change, regularly recurring energy crises, and increasing species endangerment all descending on our society like the Four Horsemen of the Apocalypse. We would do well to heed their warnings.

Appendix 1

A Guide to National Grasslands, Reservations, and Nature Preserves on the High Plains

This listing (with states and provinces sequenced alphabetically) includes more than two hundred preserved natural sites, primarily selected for being public-access locations with representative native shortgrass or mixed-grass prairies. Nearly all are situated within the major ecoregions of the Great Plains. Many smaller additional sites (generally those under 500 acres) are excluded, as are all strictly tallgrass preserves and some western sites that are largely or entirely sage-scrub or desert grassland types. The major tallgrass and other preserved sites in the eastern Great Plains and Central Lowlands have also been excluded, but these were described and mapped in an earlier work by the author (Johnsgard 2001). A more comprehensive list of North American prairie preserves, including many small sites or those outside the Great Plains, can be found in the recent book *Prairie Directory of North America* (Adelman and Schwartz 2001).

Because they are native grassland indicator species, the status of prairie dogs and burrowing owls is mentioned for those sites where such information is known to the author. The occurrence of other typical High Plains species of birds and mammals is also mentioned when known, especially for bison, pronghorns, and swift foxes. Some larger bison ranches are included in the list, as they often support other native High Plains species. The prairie dogs at sites mentioned here are all black-tailed prairie dogs, except for a few south-central Wyoming sites, where white-tailed prairie dogs occur.

ALBERTA

Alberta Public Lands. The Milk River Valley of southern Alberta has extensive publicly owned grasslands, which are used for community pastures and managed by Alberta Public Lands. Community pastures are closed to recreation. Pronghorn, elk, swift foxes (reintroduced), and greater sage-grouse are all present and may be visible from public roads or hiking trails. Several natural areas of native fescue prairies are also located in this region near the Alberta-Montana border (see the listings below).

Cypress Hills Interprovincial Park (Alberta and Saskatchewan). About 50,000 acres. The mixed-grass prairies of south-central Alberta and adjacent west-central Saskatchewan are locally known as fescue prairie, which is dominated by several species of *Festuca* and historically occurred in a crescent-shaped, northwestward-pointing arc from about Choteau, Montana, north to Edmonton, Alberta, and east to about Saskatoon and Melfort, Saskatchewan. Cypress Hills Interprovincial Park includes approximately 20,000 acres of additional disjunct fescue prairie, with the rest of the park mostly forested. No species checklists are available. Nearby (northeast of Medicine Hat) is Chappice Lake, a saline lake surrounded by grasslands, and considered one of Canada's Important Bird Areas. Also in the general vicinity of the Cypress Hills are Alkali Creek Moraine and Major Lake Prairie, both mixed-grass prairies and glacial moraine wetlands. For Cypress Hills Park information in Alberta contact the local superintendent, c/o General Delivery, Elkwater AB T0J 1C0. See also the Saskatchewan section of this appendix.

Dinosaur Provincial Park. 18,116 acres. Located north of Brooks, Alberta, with some native prairie and dinosaur fossils. The park's bird list includes 140 species, about half of which are local nesters. For information contact the park superintendent, Box 60, Patricia AB T0T 2K0.

Dry Island Buffalo Jump Provincial Park. 3,949 acres. An ancient buffalo jump on a flat-topped, grass-covered mesa rising about 600 feet above the Red Deer River. Located northwest of Drumheller, Alberta. No species checklists are available.

Little Fish Provincial Park. 151 acres. A rather small park located southeast of Drumheller, Alberta, but with nearly pristine prairie. No species checklists are available. Nearby are Little Fish Lake and Hand Hills Ecological Reserve, with extensive areas of northern fescue prairie, both considered part of Canada's Important Bird Areas.

Onefour Heritage Rangeland Natural Area. 27,589 acres. Located in three units along the Alberta-Montana border, with grasslands, badlands, and wetlands, and supporting such shortgrass species as swift fox (reintroduced), burrowing owls, mountain plovers, and ferruginous hawks, plus such mixed-grass birds as bobolinks and Baird's sparrows.

Prairie Coulees Natural Area. 4,417 acres. Dry, mixed-grass habitats along the South Saskatchewan River. Located north of Medicine Hat, Alberta.

Ross Lake Natural Area. 4,800 acres. A very large area of native fescue prairie. Located near the Alberta-Montana border, and south of Magrath, Alberta.

Rumsey Ecological and Natural Areas. About 45,000 acres. Located south of Stettler, Alberta, these two contiguous areas are a mixture of fescue grasslands, woodlands, and wetlands.

Suffield Canadian Forces Base. 660,000 acres. Located northeast of Brooks,

Alberta. A vast area of protected fescue grassland, a highly threatened community type, with probably less than 5 percent of its original acreage remaining (Winckler 2004). An area of about 113,000 acres at the base's east end has been designated as Suffield National Wildlife Area and supports many grassland species. Elk have been introduced, and pronghorns, long-billed curlews, and other grassland species can be seen from public roads around the base perimeter. Public access to the base is prohibited, but part of this large area is leased for grazing; at the west end of the base the provincial Eastern Irrigation District manages the land.

Tolman Badlands Heritage Rangeland Natural Area. About 14,500 acres. Located north of Drumheller, Alberta. Contains about 5,400 acres of fescue grasslands, with badlands that support cliff-nesting birds such as prairie falcons.

Writing-on-Stone Provincial Park. 4,245 acres. Located southeast of Milk River, Alberta. Supports one of the largest areas of protected native prairie in any of Alberta's parks. About sixty bird species are known to be local nesters. For information write Box 297, Milk River AB, TOK IMP.

COLORADO

Adobe Creek Reservoir State Wildlife Area. 5,147 acres. State-owned site managed by the Colorado Division of Wildlife, 6060 Broadway, Denver CO 80216. Located 13 miles north of Las Animas, Bent County. A reservoir with surrounding shortgrass prairie. Pronghorns, piping plovers, and golden eagles all occur. No species checklists are available.

Apishapa State Wildlife Area. 7,935 acres. State-owned site, managed by the Colorado Division of Wildlife, 6060 Broadway, Denver CO 80216. Located 32 miles east of Walsenburg, Huerfano County. Shortgrass uplands and juniper scrub. Pronghorns, mule deer, and golden eagles all occur. No species checklists are available.

Bear Creek Lake Park. 2,600 acres. Owned by the City of Lakewood, Colorado. Located at the west edge of Denver. Includes a reservoir, native grasslands, and a golf course. Prairie dogs, prairie falcons, ferruginous hawks, and golden eagles are all present. A bird species checklist of 114 species is available. Also near Denver is Barr Lake State Park, 2,600 acres of grasslands, woodlands, and wetlands, with a nature center and the headquarters of the Colorado Bird Observatory.

Bonhart Ranch. 41,000 acres. A working ranch owned by the State Land Board and managed by The Nature Conservancy. Located southeast of Colorado Springs and south of Ellicott, El Paso County. Sandsage prairie habitat in the Chico Basin, with pronghorns, prairie dogs, swift foxes, burrowing owls, mountain plovers, ferruginous hawks, and golden eagles. Not open to visitors without advance arrangements. For information contact The Nature Conservancy, 2424 Spruce St., Boulder CO 80302. Directly south of Bonhart Ranch is the 86,000-acre Chico Basin Ranch, of sandsage prairie, lakes, and mesas. It is a working ranch that is being developed for wildlife watching, hunting, and so forth. See also the entry for Pueblo Chemical Depot.

Bureau of Land Management (BLM). The BLM controls 8.3 million acres of land in Colorado (12.5 percent of the total land area), much of which is arid, upland habitat that supports shortgrass or scrub-steppe species, including Gunnison's, white-tailed, and black-tailed prairie dogs. Most of the BLM Colorado holdings lie outside the Great Plains ecoregion coverage of this book. For information on specific Colorado holdings of the BLM, contact the Bureau of Land Management, 2850 Youngfield St., Lakewood CO 80215.

Comanche National Grassland. 435,028 acres. Federally owned site managed by the U.S. Forest Service, Box 127, Springfield CO 81073. Located near Springfield and Campo, Baca County. Shortgrass and mixed-grass prairies, with some canyon topography, including Picture Canyon and Carizo Canyon. Bird species checklist are available (345 species; 72 nesters, including those of the Cimarron National Grassland, Kansas). Like the nearby Cimarron National Grassland of Kansas (see the Kansas section of this appendix), this area is critically important for supporting the declining lesser prairie-chicken and other typical sandsage and grassland birds, including Cassin's sparrow, the lark bunting, and the long-billed curlew. The area also has prairie dog colonies (about 1,400 acres in 1998) and burrowing owls. In 1998, 48 out of 129 prairie dog towns surveyed had burrowing owls. A public-access lesser prairie-chicken lek is located 14 miles east-southeast of Campo. Also in Baca County, north of Springfield, Colorado, is Two Buttes State Wildlife Area (4,962 acres). It includes shortgrass prairie around a reservoir. Prairie dogs are also present here. No species checklists are available. For information contact the Colorado Department of Wildlife, 6060 Broadway, Denver CO 80216. Near Pritchett, Colorado, is Fossil Tracks, a private 2,2180-acre preserve operated by the Southern Plains Land Trust.

Daniels Park. 1,040 acres. Part of Denver Mountain Parks. A municipal or county park that is a shortgrass prairie natural area and bison preserve, with about thirty head. Located in Douglas County, 21 miles south of Denver on Daniels Park Road, in the Colorado Rocky Mountains ecoregion of The Nature Conservancy.

Fox Ranch. 14,700 acres. A Nature Conservancy ranch on the Arikaree River in northeastern Colorado (southwest of Wray, Yuma County). Over 8 miles of river valley and mesic grasslands, supporting such eastern tallgrass species as greater prairie-chickens and grasshopper sparrows as well as typical western forms. A working ranch, with cattle present; not open to visitors without permission. For information contact The Nature Conservancy, 2424 Spruce St., Boulder CO 80302. No species checklists are available.

John Martin Reservoir State Wildlife Area. 22,000 acres. State-owned site managed by the Colorado Division of Wildlife, 6060 Broadway, Denver CO 80216. Located 5 miles east of Las Animas, Bent County. Shortgrass prairie around a large reservoir. A very large prairie dog colony is present, together with burrowing owls, golden eagles, ferruginous hawks, and many other High Plains species. No species checklists are available.

Mendano-Zapata Ranch Preserve. 100,000 acres. A Nature Conservancy preserve and working ranch, now being partly reseeded to native grasses. Previously a bison ranch (Rocky Mountain Bison Ranch). Located near Mosca, Alamosa County, in the Colorado Rocky Mountains ecoregion of The Nature Conservancy. About two thousand bison are being maintained here. For information contact The Nature Conservancy, 2424 Spruce St., Boulder CO 80302. No species checklists are available. This is the largest of several managed bison herds in Colorado. Some of the many for-profit bison ranches in Colorado are the Linnebur Grain and Buffalo Ranch, near Denver; the Denver Buffalo Company, near Kiowa, Elbert County; Rocky Mountain Bison, near Center, Saguache County; Lay Valley Bison Ranch, near Craig, Moffat County; Colorado Bison Company, near Mead, Weld County; DeVries Buffalo Ranch, near Olanthe, Montrose County; and Wolf Springs Ranch, near Westcliffe, Custer County.

Pawnee National Grassland. 693,000 acres. Federally and privately owned site managed by the U.S. Forest Service, 660 O St., Greeley CO 80631–3033. Of the total acreage 193,000 acres are federally owned; the rest is privately held. Shortgrass high plains in Weld County. There are also some isolated "Pawnee Buttes," eroded sedimentary promontories rising 350 feet, which are attractive to nesting raptors. A bird species checklist is available (284 species). Prairie dog colonies (about 730 acres in 1998) and breeding burrowing owls also occur. This is a critically important area for preserving the swift fox, mountain plover, ferruginous hawk, prairie falcon, golden eagle, and many other typical shortgrass species, and one where many significant ecological studies have been performed.

Picket Wire Canyonlands. Federally owned site managed by the U.S. Forest Service. Located 30 miles south of La Junta, just east of Piñon, Pueblo County. Near the Piñon Canyon Military Reservation (see the following entry), and along the Purgatorie River. Shortgrass birds include golden eagles, prairie falcons, ferruginous hawks, long-billed curlews, and mountain plovers. Pronghorns and swift foxes also occur. No species checklists are available.

Piñon Canyon Military Reservation. 236,000 acres. Federal site controlled and managed by the U.S. Army (Fort Carson). The reservation headquarters is located 41 miles northeast of Trinidad, Las Animas County. Public access is restricted. Shortgrass prairie and juniper scrub, along steep canyons and the Purgatorie River. Pronghorns are abundant (about a thousand head), golden eagles, ferruginous hawks, prairie falcons, and prairie dogs also occur. No species checklists are available.

Plains Conservation Center. 5,000 acres. Located off State Highway 30 and Hampton Avenue, near Aurora (a Denver suburb). Privately owned and operated. Shortgrass prairie habitat, with prairie dogs, burrowing owls, golden eagles, pronghorns ferruginous hawks, and other typical High Plains species.

Pueblo Chemical Depot. 23,000 acres. Federal site controlled and managed by the U.S. Army. Sandsage and shortgrass habitat in Pueblo County, with swift foxes, pronghorns, mountain plover, and burrowing owls. The depot is part of a 151,000-acre grassland complex that includes the Chico Basin Ranch and the Bonhart Ranch (see the Bonhart Ranch entry earlier). Visits are allowed with advance permission, except for some areas where underground ordinance or dangerous chemicals may be stored. No species checklists are available.

Rocky Mountain Arsenal National Wildlife Refuge. 17,000 acres. Federally owned refuge managed by the U.S. Fish & Wildlife Service, Bldg. 613, Commerce City CO 80032. Located in Adams County near the northeastern edge of Denver. Short grasses and wooded riparian areas; the site is locally highly polluted from past accumulation of agricultural pesticides and chemical weapons storage. Species checklists of 227 birds, 32 mammals, and 18 reptiles and amphibians are available. About 8,000 acres of shortgrass prairie are being restored. Ferruginous hawks, golden eagles, and many other arid grassland and riparian species are present. Burrowing owls and prairie dogs are also common, with the latter's colony area having declined from 4,574 acres in 1988 to 618 acres in 2001, following several sylvatic plague epidemics.

Two Buttes State Wildlife Area. 4,962 acres. State-owned site managed by the Colorado Division of Wildlife, 6060 Broadway, Denver CO 80216. Located 21 miles north of Springfield, Baca County. Shortgrass prairie around a reservoir. Pronghorns, prairie dogs, and other shortgrass species are present. No species checklists are available.

KANSAS

Big Basin Prairie Preserve. 1,818 acres. State-owned site managed by the Kansas Department of Fish & Wildlife, 808 McArtor, Dodge City KS 67801. Located in Clark County, about 16 miles south of Minneola. Mixed-grass prairie, with bison. Nearby is Clark County Lake and Wildlife Management Area (1,240 acres), with mixed-grass prairie along Bluff Creek and an associated canyon. No species checklists are available for either area.

Cedar Bluff Reservoir and Wildlife Area. 17,500 acres. Federally owned (Bureau of Reclamation) site managed by the Kansas Department of Wildlife & Parks, 808 McArtor, Dodge City KS 67801. Located in Trego County, southeast of Wakeeney. Limestone bluffs, upland prairies, and woodlands, including a designated wildlife area at the reservoir's western end. Partly mixed-grass prairie, with mule deer, black-tailed jackrabbits, and coyotes, plus ferruginous hawks and prairie falcons in winter. Prairie dogs may also be present. No species checklists are available.

Cheyenne Bottoms Waterfowl Management Area. 18,000 acres. State-owned site managed by the Kansas Department of Fish & Wildlife, Rte. 1, Great Bend KS 67530. Marshes and moist, mixed-grass prairie, managed mainly for shorebirds and waterfowl. A bird list is available (319 species; 104 nesters); this is one of the most important migratory stopover points for shorebirds and waterfowl in the Great Plains. Northwest of the waterfowl management area (at the corner of E. 100th Rd. and N. 20th Ave.) is a small adjoining Nature Conservancy preserve, with a prairie dog town and resident burrowing owls.

Cimarron National Grassland. 108,175 acres. Federally owned site managed by the U.S. Forest Service, 242 E. Highway 56, Box 300, Elkhart KS 67950. Shortgrass plains and sandsage steppe. A bird species checklist is available (including the Comanche National Grassland) of 345 species, with 72 nesters. Cable, Seltman, and Cook (1996) provided species checklists of 47 mammals, including the swift fox and the pronghorn, as well as 22 reptiles and 9 amphibians. There are many prairie dog colonies (about 1,300 acres in 1998), and burrowing owls are common nesters. Most (31 of 38) prairie dog towns surveyed had burrowing owls in 1998. Public-access blinds at lesser prairie-chicken leks are available. This grassland is also an important habitat for lark buntings, Cassin's sparrows, Brewer's sparrows, and other sandsage-associate species. Mountain plovers are rare breeders.

Finney Game Refuge. 3,800 acres. State-owned site managed by the Kansas Department of Wildlife & Parks, 808 McArtor, Dodge City KS 67801. Located in Finney County, 1 mile south of Garden City, Finney County. Shortgrass and sandsage prairie. A small colony of prairie dogs is present, and more than one hundred bison are maintained in confined pastures. State-owned bison herds are also maintained at the 2,500-acre Maxwell Game Reserve, near Canton (McPherson County), with about two hundred head, and on the 4,530-acre Byron Walker Wildlife Area, near Kingman (Kingman County), both located in south-central Kansas.

Gypsum Hills Scenic Drive. The Gypsum Hills (or "Red Hills") of southwestern Kansas are privately owned, but a drive through them takes one through scenic mixed-grass prairie and cedar woods, with coyotes, prairie dogs, burrowing owls, and many other grassland or scrub-adapted bird species, especially in winter and during migration. The Gypsum Hills Scenic Drive is a 20-mile loop, starting and ending a few miles west of Medicine Lodge. Also in southwestern Kansas (near Stubbs, Barber County) is Ted Turner's Z-Bar Ranch (42,479 acres), with a managed herd of about 1,800 bison.

Kanopolis State Park. 18,000 acres. Federally owned site surrounding a Corps of Engineers reservoir, managed as a park by the Kansas Department of Wildlife & Parks. Located near Marquette, about 30 miles southwest of Salina, Saline County. A reservoir in the Smoky Hill River Valley of the Dakota Hills, with a nature trail through diverse habitats. Prairie dogs, mule deer, horned larks, western meadowlarks, and other grassland birds are present in the mixed-grass prairie habitats of this valley. No species checklists are available.

Kirwin National Wildlife Refuge. 10,778 acres. Federally owned refuge. Manager's address: Rte. 1, Box 103, Kirwin KS 67644. Mixed-grass prairie and hardwoods surrounding the 5,000-acre Kirwin Reservoir. A bird species checklist is available (197 species, 46 nesters) as well as lists of 34 mammals and 31 species of reptiles and amphibians. Prairie dogs are present (60 acres in 1994), and burrowing owls are occasional nesters.

Logan Wildlife Area. 271 acres. State-owned site managed by the Kansas Department of Wildlife & Parks, 808 McArtor, Dodge City KS 67801. Located 1.5 miles north and 2.5 miles west of Russell Springs, Logan County. Contains shortgrass habitats. Prairie dogs and burrowing owls are fairly common in the general vicinity of Logan County. The steep-sided Chalk Pyramids (also called "Monument Rocks"), located on private land east of Highway 83 between Oakley and Scott City, are eroded remnants of the ancient Cretaceous-era seabed and provide interesting examples of High Plains topography in the paleontologically rich Smoky Hill Valley. In the general vicinity are pronghorns, ferruginous hawks, and golden eagles, plus prairie falcons in winter. Farther east, driving south on Castle Rock Road from Quinter takes one to the similar Castle Rock outcrop (also on private land), and there is a return route back on Banner Road to Collyer. Pronghorns, black-tailed jackrabbits, prairie dogs, and similar shortgrass wildlife species all occur in this region.

Lovewell State Park and Reservoir. 6,275 acres. Federally owned (Bureau of Reclamation) site managed by the Kansas Department of Wildlife & Parks, R.R. 1, Box 666, Webber KS 66970. Upland shortgrass prairies, chalk bluffs, and oak woodlands around a reservoir in Jewell County, located northeast of Mankato. A prairie dog colony is present, along with the associated grassland and woodland species. No species checklists are available.

Mount Sunflower. Located about 30 miles northwest of Sharon Springs, Wallace County. On private land, this highest point in Kansas (4,039 feet) is almost on the Colorado border, in shortgrass prairie. Swift foxes, coyotes, pronghorns, and several western raptors are sometimes visible from the roadsides. Directly north, about 25 miles northwest of St. Francis and along the Colorado and Nebraska borders, are the Arikaree River Breaks. Also on private land, the shortgrass prairies here support prairie dogs, coyotes, mule deer, prairie dogs, and burrowing owls. A loop route taking one about 2.5 miles north of the river valley (on the Nebraska side of the river) provides the most scenic viewing.

Prairie Dog Lake State Park and the Norton Wildlife Area. 7,850 acres. State-owned site managed by the Kansas Department of Wildlife & Parks, Box 338, Hays KS 6701–0338. Located in Norton County, 5 miles southwest of Norton. Mixed-grass prairie and an impoundment of Prairie Dog Creek. Prairie dogs (a colony of about 38 acres in 2000) and burrowing owls are present. No species checklists are available.

Quivira National Wildlife Refuge. 21,820 acres. Federally owned refuge. Manager's address: Box G, Stafford KS 67578. Mature hardwoods, sandhill grasslands, rangeland, farmland, and alkaline marshes. A bird species checklist is available (252 species, 88 nesters). Prairie dogs are present, and burrowing owls are common nesters.

Sandhills State Park. 1,123 acres. State-owned site managed by the Kansas Department of Wildlife & Parks, 808 McArtor, Dodge City KS 67801. Located at the northeast edge of Hutchinson, Reno County. Sandsage grasslands along the Arkansas River, with associated species. No species checklists are available. Prairie dogs have not been reported here; it may be too sandy for them.

Scott State Park. 1,120 acres. State-owned site managed by the Kansas Department of Wildlife & Parks, 808 McArtor, Dodge City KS 67801. Located 13 miles north-northwest of Scott City, Scott County. Managed herds of elk and bison on High Plains grasslands, with steep canyons along Ladder Creek, a nature trail, and such associated species as rock wrens and prairie falcons. No species checklists are available. Also near Scott City is the privately owned Duff Land and Cattle Company, with a managed bison herd of nearly one thousand animals. Other for-profit bison ranches in Kansas include Butterfield's Buffalo Ranch, near Beloit, Mitchell County, and Flying G Buffalo, near Codell, Rooks County. Fort Riley Military Reservation, Riley County, also maintains a buffalo corral.

Smoky Valley Ranch. 16,800 acres. A Nature Conservancy preserve in Logan County that contains shortgrass prairie, chalk bluffs, rocky ravines, and other upland habitats. Located 18 miles south and 5 miles west of Oakley, Logan County. Supports about 150 bison as well as pronghorns, prairie dogs, burrowing owls, golden eagles, and other shortgrass species. For information contact The Nature Conservancy, 700 sw Jackson, Suite 804, Topeka ks 66612.

Webster State Park. 10,380 acres. Federally owned reservoir site (Bureau of Reclamation) managed by the Kansas Department of Wildlife & Parks, 808 McArtor, Dodge City ks 67801. Located 8 miles west of Stockton, Rooks County. Partly mixed-grass prairie around a reservoir, with mule deer, coyotes, western meadowlarks, magpies, and other grassland species, especially along the Webster Wildlife Tail, which follows the Solomon River for several miles. No species checklists are available.

Wilson State Park and Rocktown Natural Area. 10,086 acres. Federally owned reservoir site (Corps of Engineers) managed by the Kansas Department of Wildlife & Parks, 808 McArtor, Dodge City ks 67801. Located in Russell County, southwest of Sylvan Grove. Mixed-grass prairie is best preserved within the Rocktown Natural Area of the park. An effort is being made to reestablish golden eagles here. No species checklists are available.

MONTANA

Bighorn Canyon National Recreation Area. 63,000 acres. Federally owned site located along the Montana-Wyoming border in Bighorn County, adjacent to the Crow Indian Reservation. Manager's address: P.P. Box 458, Fort Smith mt 59035. Canyons and prairies, with grassland and brush-adapted wildlife.

Benton Lake National Wildlife Refuge. 12,383 acres. Federally owned refuge. Manager's address: 922 Bootlegger Trail, Great Falls mt 59404. Located 14 miles north of Great Falls. Extensive native (mostly shortgrass) prairie (about 6,000 acres), with sharp-tailed grouse (viewing blind available), long-billed curlew, marbled godwit, burrowing owl, Sprague's pipit, McCown's and chestnut-collared longspurs, Baird's sparrow, and other shortgrass and mixed-grass species. A vertebrate species checklist is available (243 bird species, 73 known nesters). There are also twenty-eight reported mammals, eight reptiles, and two amphibians. White-tailed jackrabbits and coyotes are common.

Bowdoin National Wildlife Refuge. 15,437 acres. Federally owned refuge. Manager's address: Box J, Malta mt 59538. Native shortgrass and mixed-grass prairies (about 7,000 acres) and wetlands, with many waterbirds and shorebirds. A bird species checklist is available (263 species, 102 nesters). Burrowing owls, ferruginous hawks, and long-billed curlews are all uncommon breeders. Sprague's pipits, Baird's sparrows, lark buntings, and chestnut-collared longspurs also breed, and pronghorns are common.

BR-12 *Prairie Marsh.* 1,800 acres Federally owned site managed by the Bureau of Land Management, Havre Resource Area, Havre MT 59501. Located about 10 miles north of Zurich, Blaine County. Surrounds a 200-acre marsh (the impounded Fifteenmile Creek) in native prairie, with ferruginous hawks, golden eagles, and other prairie wildlife.

Bureau of Land Management (BLM). The BLM controls 8.1 million acres of federally owned, public-access lands in Montana (8.7 percent of the total land area). Several important grasslands occur within Extensive Recreation Management Areas (ERMAS). Those that are known to have supported prairie dogs in the 1980s or 1990s and therefore probably have extensive grasslands are Big Dry ERMA (1,422,350 acres, 7,360 prairie dog acres in 1991), Phillips ERMA (740,690 acres, 5,923 prairie dog acres in 1993), Powder River ERMA (1,040,175 acres, 3,200 prairie dog acres in 1986), Billings ERMA (310,590 acres, 1,300 prairie dog acres in 1986), Havre ERMA (282,000 acres, 1,000 prairie dog acres in 1994), and Valley ERMA (366,486 acres, 830 prairie dog acres in 1994). Second to the Indian reservations, the BLM is probably the largest holder of lands occupied by prairie dogs in the state. Prairie dogs are considered a sensitive species by the BLM, which has established locally variable target acreages for prairie dogs. No species checklists are available. For locations contact the Bureau of Land Management, 222 N. 32nd St., Billings MT 59107. There are also regional offices in Lewistown and Miles City.

Charles M. Russell National Wildlife Refuge. 1,094,000 acres. Federally owned refuge. Manager's address: Box 110, Lewistown MT 59457. Shortgrass prairie (about 97,000 acres), brush, and shoreline hardwoods around Fort Peck Reservoir (about 245,000 acres). Much of this vast refuge is essentially still wilderness and largely inaccessible by car, as is also true of the contiguous UL Bend NWR and the Upper Missouri River National Monument (see the entry for the Upper Missouri River Wild and Scenic River). A bird species checklist is available (252 species, 98 nesters). Pronghorns, elk (reintroduced in 1951), bighorns (reintroduced in 1947), and deer are present; the refuge's elk population is the nation's largest herd of prairie elk (nearly four thousand animals in the mid-1990's). Prairie dogs are common (about 6,500 colony acres in 1994); Manning Corral Prairie Dog Town is located along the northern border of the refuge, south of U.S. Highway 199, about where (undeveloped) state route 202 enters the refuge. This prairie dog colony was seriously affected by sylvatic plague in 1992 and is slowly recovering. Black-footed ferrets were reintroduced here starting in 1994, and in 1996 an insecticide dusting program was started to prevent the spread of sylvatic plague. Pronghorns, elk, mountain plovers, and other shortgrass species are common here. The Pines Recreation Area is located about 30 miles southwest of Fort Peck, on Pines Road along the north shore of Fort Peck Lake.

High Plains grasslands and pines, with golden eagles, prairie falcons, ferruginous hawks, marbled godwits, greater sage-grouse, and other shortgrass and sage-steppe species present.

Comertown Pothole Prairie Preserve. 1,130 acres. A Nature Conservancy preserve of native mixed grasses and shallow glacial "potholes," similar to the habitats of nearby Medicine Lake National Wildlife Refuge. Located 15 miles northeast of Plentywood, in Sheridan County. For information contact The Nature Conservancy, 32 S. Ewing, Helena MT 59601.

Crow Indian Reservation. About 1,500,000 acres, of which about 400,000 are tribally owned. Headquarters: Crow Agency MT 59022. The 2000 Native American population living on the reservation was about 7,500. In 1994 there were 2,000 acres of prairie dog colonies present. This reservation probably has extensive areas of native grassland plants and animals, but no faunal surveys are available. A herd of about 1,200 bison is maintained.

Crown Butte Preserve. A Nature Conservancy preserve, with several mixed-grass and bunchgrass habitat types present. Located 30 miles southwest of Great Falls, Cascade County, within the Canadian Rocky Mountains ecoregion of The Nature Conservancy. For information contact The Nature Conservancy, 32 S. Ewing, Helena MT 59601.

Flying D Ranch. 113,593 acres. A Ted Turner ranch near Gallatin Gateway, Gallatin County, holding over two thousand bison and supporting other native shortgrass wildlife. Other large bison ranches in Montana include Buffalo Jump Ranch, Bozeman, Gallatin County; Carter Ranch, near Livingston, Park County; and the Blackfeet Indian Reservation, near Browning, Glacier County.

Fort Belknap Indian Reservation. 645,575 acres. Federal Indian reservation. Headquarters: Fort Belknap Agency MT 59526. This large Gros Ventre and Assiniboine reservation has a managed bison herd (about four hundred head on a 10,000-acre reserve), prairie dog colonies, burrowing owls, mountain plovers, and (as of 1997) reintroduced black-footed ferrets. In 1994 there were an estimated 20,614 acres of prairie dogs, the largest acreage known for any Montana site. The reservation has closed its prairie dog colonies in the ferret reintroduction site to shooting. Bison can be viewed only by guided tours. A bird list of 115 species has been developed. The 1990 Native American population was 2,332. To the west is Rocky Boy Chippewa-Cree Indian Reservation (108,015 acres). The 2000 Native American population was about 3,000. Headquarters: Box Elder, Montana. Still farther west, at the base of the Rocky Mountains, is the Blackfeet Indian Reservation, of about 1.5 million acres and with a 2000 population of 14,000 Native Americans, with about 7,000 living on the reservation. Headquarters: Browning MT 59417. Wildlife on these reservations is probably similar to that reported for the Charles M. Russell NWR but is still essentially undocumented. An Intertribal Prairie Ecosystem Restoration Consortium has recently been formed among eight northern tribes for managing their grasslands and prairie wildlife. In 2003 the total tribal suitable habitat for prairie dogs in Montana was estimated at 90,000 acres (Tim Vosburgh, Tribal Black-tailed Prairie Dog Coordinator, personal communication via Robert Luce). Also in north-central Montana, nonreservation lands in southern Phillips County had about 24,000 acres of prairie dog colonies between 1984 and 1989. Other very large colony acreages occurred in southern Custer County (about 16,500 acres), extreme northern Custer and southern Prairie County (about 6,000 acres), the Tongue River area (about 6,000 acres), northern Carter County (about 5,000 acres), and northeastern Garfield County (about 4,500 acres).

Fort Peck Indian Reservation. About 2 million acres. (of which about 250,000 are tribally owned by Native Americans). Federal Indian reservation. Headquarters: Poplar MT 59255. Located to the northeast of Charles M. Russell NWR and probably has a similar biota to that of the refuge. It is one of the two largest Indian reservations in Montana, but no information on prairie dog abundance or other wildlife is available. Indian reservations are open to the public, but tribal permits are needed for some sporting activities. The 2000 Native American reservation population (Assiniboine, Upper Yanktoni, and Sisseton-Wahpeton) was about 6,000, with another 3,800 living off the reservation.

Fox Lake Wildlife Management Area. 1,534 acres. State-owned and managed by Montana Fish, Wildlife & Parks, P.O. Box 20070, 1420 E. Sixth Ave., Helena MT 59620. Located about 2 miles southeast of Lambert, Richland County. A large prairie marsh, with waterfowl, shorebirds, and various High Plains wildlife, including pronghorns, burrowing owls, Sprague's pipits, and marbled godwits.

Greycliff Prairie Dog Town State Park and Wildlife Management Area. 98 acres. State-owned and managed by Montana Fish, Wildlife & Parks, P.O. Box 20070, 1420 E. Sixth Ave., Helena MT 59620. Located in Sweetgrass County about 8 miles east of Big Timber. A large prairie dog town with associated shortgrass wildlife, including golden eagles and burrowing owls.

Hailstone National Wildlife Refuge. 1,913 acres. Federally owned refuge managed by the Charles M. Russell NWR. Located about 5 miles northeast of Rapelje, Stillwater County. A large alkaline lake surrounded by shortgrass prairie (part of Montana's Big Open region), with a prairie dog colony (80 acres in 1994), burrowing owls, golden eagles, pronghorns, and other shortgrass wildlife. No species checklists are available.

Lake Mason National Wildlife Refuge. 16,830 acres. Located in Musselshell County, in thirty units, all from 8 to 26 miles north of Roundup (part of Montana's Big Open region). The manager's office is in Roundup. The middle segment (Willow Creek Unit) has extensive shortgrass prairie, with prairie dogs, burrowing owls, mountain plovers, and pronghorns. The North Unit is the largest, at 5,323 acres. No species checklists are available.

Little Bighorn Battlefield National Monument. 765 acres. Located 15 miles south of Hardin, Big Horn County. Federally owned historical site. Address: P.O. Box 39, Crow Agency MT 59022. Mostly pristine mixed-grass prairie, the site of Custer's fatal encounter.

Makoshika State Park. 8,832 acres. Near Glendive, Dawson County, and Medicine Rocks State Park (330 acres), near Eklaka, Carter County. Both of these state-owned parks are mostly shortgrass prairies with eroded sandstone cliffs and fossil-rich Mesozoic-era badlands. No species checklist is available, but pronghorns, mule deer, sharp-tailed grouse, and other shortgrass species are present at both. For more information contact Medicine Rocks State Park, Box 1630, Miles City MT 59301; or Makoshika State Park, Box 1242, Glendive MT 59330. Or contact the Montana State Park headquarters, 1420 E. Sixth Ave., Helena MT 59620.

Matador Ranch. 60,000 acres. A Nature Conservancy site used in part as a functioning cattle ranch. Located in Phillips County, 40 miles south of Malta. Mostly mixed-grass prairie, sage grasslands, and some streams. Prairie dogs occur here, and black-footed ferrets have been released. This general area south of Malta may support the best remaining assemblage of typical shortgrass species anywhere in the Great Plains. For information contact The Nature Conservancy, 32 S. Ewing, Helena MT 59601.

Medicine Lake National Wildlife Refuge. 31,457 acres. Federally owned refuge. Manager's address: HC51, Box 2, Medicine Lake MT 59247. Shortgrass–mixed-grass prairie transition vegetation (about 17,000 acres) and wetlands, with breeding upland sandpipers and lark buntings, as well as McCown's and chestnut-collared longspurs. Mixed-grass species, such as Sprague's pipits and LeConte's and Baird's sparrows, are also common. Burrowing owls are uncommon breeders. A bird species checklist is available (224 species, 104 nesters). North of Medicine Lake, between Westby and the Canadian border, is an area of mixed-grass prairie and potholes rich in grassland and shoreline birds, including such rare species as yellow rail, LeConte's sparrow, and Nelson's sharp-tailed sparrow.

National Bison Range. 18,542 acres. Federally owned refuge located within the Canadian Rocky Mountains ecoregion of The Nature Conservancy. Manager's address: Moiese MT 59824. Palouse bunchgrass prairies (about 15,000 acres) and scattered woods in the Flathead Valley of Lake County at 4,500 feet elevation. Primarily managed for bison (three to five hundred animals), elk, bighorn sheep, pronghorns, and deer. A bird species checklist is available (268 species,

104 nesters). Burrowing owls are occasional to rare summer visitors; golden eagles and sharp-tailed grouse are common.

Northern Cheyenne Indian Reservation. 445,000 acres. Federal Indian reservation. Headquarters: Lame Deer MT 59043. In 1994 there were 1,024 acres of prairie dog colonies present and a bison herd of about a hundred head was being developed. In 1999 prairie dogs were reintroduced to replace those lost to a plague outbreak. There are plans to reintroduce ferrets when the prairie dog populations have sufficiently recovered. The 2000 Native American population was about 5,000, mostly Northern Cheyenne. This reservation is bordered on the west by the larger Crow Indian Reservation (see earlier entry).

Pine Butte Swamp Preserve. 18,000 acres. A Nature Conservancy preserve 25 miles northwest of Choteau in Teton County that contains extensive native shortgrass prairies as well as a wetland fen, forests, and various montane habitats. Located in the Canadian Rocky Mountains ecoregion of The Nature Conservancy. Elevations range from 4,500 to 8,500 feet. Over 40 mammal species, including grizzly bears, and 150 bird species, including such typical grassland birds as long-billed curlew, sharp-tailed grouse, and LeConte's sparrow, have been reported. For information contact the Pine Butte Guest Ranch, HC58 Box 34C, Choteau MT 59422. No species checklist is available.

Rosebud Battlefield State Park. 3,052 acres. A state-owned historic park that is the site of a battle that occurred eight days before Custer's final encounter with the Cheyenne and Dakota Sioux. Native shortgrass prairie. Located 24 miles south of Busby, Rosebud County.

South Ranch Conservation Easement. 151,978 acres. Also Tampico Conservation Easements (3,770 acres) and Bitter Creek Wilderness Study Area (59,600 acres). Privately owned, state-owned, or federally owned lands in conservation easements or identified as Areas of Critical Environmental Concern (ACEC). Shortgrass prairie and sage grasslands, with prairie dogs, mule deer, prairie falcons, and other shortgrass species. Bitter Creek Wilderness Study Area is located about 40 miles north of Hinsdale, Valley County, along Rock Creek Road. South Ranch Conservation Easement is 35 miles southwest of Glasgow, Valley County, and is managed by Montana Fish, Wildlife & Parks. Montana Fish, Wildlife & Parks also manages the Tampico Conservation Easements, which are 11 miles west of Glasgow. For information on the last two areas contact the state's ACEC office at 1430 E. Sixth St., Helena MT 59620.

Terry Badlands Special Recreation Management Area. 43,000 acres. Federally owned site managed by the Bureau of Land Management, P.O. Box 36800, Billings MT 59107–6800. Located about 8 miles northwest of Terry, Prairie County. A badlands area along the Yellowstone River, with a mixture of native grasses, scrub woodland, and conifers. Pronghorns, golden eagles, prairie falcons, ferruginous hawks, long-billed curlews, mountain plovers, greater sage-grouse, and Sprague's pipits all occur here. Prairie dogs are present, and swift foxes have been reported.

UL *Bend National Wildlife Refuge (and associated* UL *Bend Wilderness Area).* 50,049 acres. Federally owned refuge. This refuge (of which 20,819 acres are a designated wilderness area) is even more remote than the Charles M. Russell NWR, with which it is contiguous and from which it is managed (contact: Box 110, Lewistown MT 59457. A bird species checklist (252 species, 98 nesters) encompassing both refuges is available. Burrowing owls are common breeders. Other common High Plains species present are sharp-tailed grouse, greater sage-grouse, long-billed curlews, and golden eagles. Mountain plovers are uncommon nesters. There were about 5,500 acres of prairie dog colonies in the Russell and UL Bend refuges as of 2000. Pronghorns, elk, and deer are present on both refuges.

Ulm Pishkun State Park. 170 acres. A state-owned park managed by Montana Fish, Wildlife & Parks, 1420 E. Sixth St., Billings MT 59101. Located at the westernmost edge of the Great Plains, with a mile-long "piskun," a place where, historically, bison were stampeded into jumping from steep cliffs. This site is probably the largest reported such bison-killing site in the United States. Some native prairie has been preserved. A prairie dog town is present, and pronghorns are common, as are prairie rattlesnakes. Located near Ulm, Cascade County, about 5 miles west of the Great Falls airport.

Upper Missouri National Wild and Scenic River. 92,000 acres. Federal site managed by the Bureau of Land Management (BLM), Airport Rd., Lewistown MT 59457, and the associated Upper Missouri River National Monument (377,00 acres of BLM land, plus 39,000 acres of state land and 80,000 acres of private land). A 149-mile section of the Missouri River, extending from near Fort Benton to Kipp State Park (U.S. Highway 191) and the westernmost portion of the Charles M. Russell NWR. This stretch of the upper Missouri River was officially designated as a national wild and scenic river in 1973, and the surrounding lands were named a national monument in 1999 by presidential decree. However, local political pressures may endanger this important preservation status, even though prior mining, gas-drilling, and grazing rights will largely be continued. The terrestrial vertebrates of the currently designated national monument area include 233 birds, 60 mammals, and 20 reptiles and amphibians, many of which are the same as those reported for the Charles M. Russell NWR. At Holmes Rapids there is a 200-acre prairie dog town, and prairie falcons are regular. Historically famous and spectacular castlelike white sandstone formations occur from below the mouth of the Marias River east for about 40 air miles (55 river miles), or to about Arrow Creek, but are not visible by normal road access. The Missouri Breaks National Back Country Byway is an 81-mile unimproved road loop starting and ending 10 miles east of Winfred, off state route 236.

Agate Fossil Beds National Monument. 3,150 acres. Federally owned site. Manager's address: Box 427, Gering NE 69341. Shortgrass plains with extensive Miocene-age fossil deposits, located a few miles east of Agate, Sioux County. No vertebrate species checklists are yet published, but a preliminary bird list comprises 156 species (including burrowing owl, golden eagle, and prairie falcon). Thirty mammals (including swift fox and pronghorn) have also been reported.

Cherry Ranch. 7,260 acres. Located 10 miles south of Harrison in Sioux County. Owned by The Nature Conservancy and leased in part for cattle ranching. Mixed-grass and shortgrass prairie with bluffs and rock outcrops, much like the nearby Agate Fossil Beds National Monument (see entry) and the Guadalcanal Ranch (see entry) in its ecology and associated biota. For information contact The Nature Conservancy, 2029 Leavenworth, Suite 100, Omaha NE 68102.

Chimney Rock National Historic Site. 83 acres. Federally owned site managed through Scotts Bluff National Monument (see entry). Manager's address: Box F, Bayard NE 59334. Located 4 miles south of Bayard, Morrill County. Largely undeveloped. This famous 425-foot, spire-shaped landmark of clay, volcanic ash, and sandstone and the nearby Jail Rock and Courthouse Rock (both 4 miles south of Bridgeport) are historic sites that were important monolithic markers along the Oregon Trail. Golden eagles have nested on Chimney Rock and Jail Rock, and burrowing owls have regularly occurred in a prairie dog colony south of Chimney Rock. No species checklist is available.

Crescent Lake National Wildlife Refuge. 45,849 acres. Federally owned refuge. Manager's address: Star Rte., Ellsworth NE 69340. Sandhills grasslands (about 37,000 acres) and dozens of small, mostly alkaline shallow lakes and marshes. A bird species checklist is available (279 species, 86 nesters). Sharp-tailed grouse, long-billed curlews, and upland plovers are common breeders. Burrowing owls are uncommon nesters, but the sandy substrate is generally unsuitable for prairie dog burrows. Introduction of bison is under consideration. A large Ted Turner ranch (Blue Creek Ranch) that is primarily used for bison rearing is located between Oshkosh and Crescent Lake. One of the ranch's conservation projects involves prairie dog restoration.

Fort Niobrara National Wildlife Refuge. 19,131 acres. Federally owned refuge. Manager's address: Hidden Timber Rte., Valentine NE 69201. Riverine woods, mixed-grass uplands, and sandhills prairie, managed for bison (about 500), elk (40), and other wildlife (275). Pronghorns are occasionally present, and seven species of bats have been reported, as well as both species of jackrabbits. There is a prairie dog colony (55 acres in 1994) and burrowing owls near the refuge headquarters. Burrowing owls are common nesters, as are sharp-tailed grouse, northern harriers, and Swainson's hawks. A bird species checklist is available (20l species, 76 nesters). To the east (about 30 miles east of Valentine) is The Nature Conservancy's Niobrara River Valley Preserve (see entry).

Fort Robinson State Park. 22,000 acres. State-owned site managed by the Nebraska Game and Parks Commission, P.O. Box 392, Crawford NE 69339. Located a few miles west of Crawford, Dawes County. A mixture of shortgrass prairie (about 50 percent) and ponderosa pine savanna, with some high bluff topography. Pronghorns, bighorns, mule deer, and related shortgrass and woodland species occur here, as well as bison (about three to four hundred head). The adjacent state-owned Peterson Wildlife Management Area (22,640 acres) is similar, as is the nearby federally owned Soldier Creek Wilderness (9,600 acres). The Soldier Creek Wilderness area was extensively burned in 1989. It must be entered through Fort Robinson State Park.

Guadalcanal Ranch. 5,000 acres. A working ranch located about 6 miles southwest of Harrison, Sioux County. Mostly consists of shortgrass prairie. Owned and managed by the Prairie Plains Resource Institute, Aurora NE 68818. Permission from the Institute is required for entry. Located a short distance northwest of The Nature Conservancy's Cherry Ranch (see entry above). McCown's and chestnut-collared longspurs, long-billed curlews, prairie falcons, golden eagles, and ferruginous hawks are among the local breeding birds. West of Harrison a county road going south from U.S. 20 that is just east of the Wyoming state line passes through wonderful shortgrass prairie and all the usual shortgrass species. Five miles north of Harrison is Gilbert Baker Wildlife Management Area, which has relatively little grassland and is mostly riparian woodland and pine forest.

Hutton Niobrara Ranch Wildlife Sanctuary. 4,500 acres. A ranch recently (2003) bequeathed to Audubon of Kansas and still in its formative stages as a sanctuary. Located between Bassett, Rock County, and the Niobrara River, with many of the same sandhill prairie and riverine habitats as the nearby Niobrara River Valley Preserve (see entry below).

Lillian Annette Rowe Sanctuary. About 2,000 acres. A sanctuary on the Platte River in Buffalo County. Located about 2 miles south and two miles west of Gibbon, Buffalo Co., Nebraska. Owned and managed by the National Audubon Society, Rte. 2, Box 146, Gibbon NE 68840. Primarily a bird sanctuary, especially established for migrating cranes but with about 300 acres of native prairie (mostly wet meadows and taller grasses). No species checklists are available.

Downstream to the east as far as Grand Island are several parcels of similarly subirrigated grasslands along the Platte River that are mostly owned either by the Nature Conservancy or by the Platte River Whooping Crane Habitat Management Trust (2,500 acres), 6611 Whooping Crane Dr., Wood River NE 68883. These lands are valuable resting and foraging areas for cranes and geese.

Nebraska National Forest. 360,000 acres. Federally owned sites, consisting of three rather widely separated U.S. Forest Service units. Headquarters: 125 N. Main St., Chadron NE 69337. A bird species checklist is available from the Chadron office for the entire Pine Ridge region (about 250 species, 36 nesters). Burrowing owls are uncommon breeders, as are golden eagles, ferruginous hawks, and prairie falcons The southernmost Bessey District (90,448 acres) is mostly sandhills prairie, with extensive stands of planted conifers (the "forest") and a local bird species checklist of 95 species (36 nesters). Prairie dog colonies (about 70 acres in 1998) and burrowing owls are present along the Dismal River. Sharp-tailed grouse are common, and greater prairie-chickens are also present. The Samuel R. McElvie Forest (115,703 acres) to the north (26 miles southwest of Valentine, Cherry County) is mostly composed of sandhills grassland, with some pine plantings and no reported prairie dogs or burrowing owls. A sharp-tailed grouse blind is available for public use. The Pine Ridge District (50,803 acres) of the Nebraska National Forest is near Chadron, Dawes County, and is a mixture of shortgrass prairie and native ponderosa pine forest. The state-owned Metcalf Wildlife Management Area (3,068 acres) is located to the east of Chadron near Hay Springs, Sheridan County. It has canyon-and-bluff habitats with some mixed-grass prairies as well as pine woodlands, and its resident birds include prairie falcons and golden eagles.

Niobrara River Valley Preserve. About 63,000 acres. A nature preserve owned and managed by The Nature Conservancy, 1019 Leavenworth, Suite 100, Omaha NE 68102. Located along the Niobrara River, in eastern Cherry and Brown counties. The headquarters is about 15 miles north of Johnstown, Brown County. Riparian transition zone (of about 25 river miles) between western ponderosa pine and eastern hardwood floodplain forest, with adjoining sandhills prairie uplands. Lies east of Fort Niobrara National Wildlife Refuge (see entry above), within the Niobrara Scenic River District. Partly managed for bison (250 to 400 head). Prairie dogs and burrowing owls are both present. Local species checklists include 213 birds, 44 mammals, 17 reptiles, 8 amphibians, 25 fish, 70 butterflies, 86 lichens, and 581 vascular plants. A total of 105 presumptive breeding bird species has been documented. These include several pairs of closely related eastern and western forms (towhees, bluebirds, wood-pewees, grosbeaks, orioles, buntings), as well as occasional hybrid combinations of these species pairs that breed within this unique ecologic and zoogeographic transition zone.

Oglala National Grassland. 94,344 acres. Federally owned site managed by the U.S. Forest Service, 16524 Highway 385, Chadron NE 69337. Shortgrass prairie and eroded badlands. Typical shortgrass fauna, including such High Plains raptors as the golden eagle and the ferruginous hawk. No bird species check-list specifically limited to the Oglala National Grassland is available, but a list of 302 species that encompasses the Forest Service's entire Pine Ridge Recreational Area of northwestern Nebraska is available. Long-billed curlews and upland sandpipers are common, ferruginous hawks and golden eagles are regular nesters, and both longspur species breed commonly. Swift foxes have recently been reported. There are prairie dog colonies (about 740 acres in 1998), but burrowing owls are uncommon. Toadstool Geological Park (about 300 acres) is an undeveloped, federally owned badlands area (located 19 miles northwest of Crawford, Dawes County) that has Oligocene-age exposures and is situated within the national grassland boundaries. Little vegetation or surface water is present, but arid-adapted species, such as snakes and rock wrens, are abundant, and such raptors as prairie falcons and golden eagles use the bluffs for nesting. The state-owned Ponderosa Wildlife Management Area (3,660 acres) is also near Crawford. Like the Metcalf Wildlife Management Area (see the Nebraska National Forest entry above), it has canyon-and-bluff habitats with some mixed-grass prairies and ponderosa pine woodlands.

Scotts Bluff National Monument. 3,100 acres. Federally owned historic site managed by the National Park Service. Manager's address: Box 27, Gering NE 69341. A pine-covered, 800-foot bluff surrounded by shortgrass uplands; an important site on the Oregon Trail. A terrestrial vertebrate survey documented 98 bird species, 28 mammals, and 12 reptiles and amphibians. Prairie dogs (2 acres in 1994) and burrowing owls are present. Several Nebraska birds otherwise found only in the ponderosa pine forests of the state's northwestern Pine Ridge, such as dark-eyed juncos and violet-green swallows, breed here, as do some cliff nesters, such as white-throated swifts and prairie falcons. About 10 miles south of Gering, Scotts Bluff County, is the state-owned Wildcat Hills State Recreation Area (335 acres) and the adjacent Buffalo Creek Wildlife Management Area (2,880 acres), with many of the same bluff-and-canyon habitats and species found at Scotts Bluff and in the Pine Ridge region. Golden eagles and prairie falcons are present, and common poorwills are conspicuously prevalent in summer. A few bison are also maintained here. There is no bird list, but an interpretive center is present at the Wildcat Hills SRA and can provide local information.

Valentine National Wildlife Refuge. 71,516 acres. Federally owned refuge managed by the Fort Niobrara NWR, Hidden Timber Rte., Valentine NE 69201. Marshes, dozens of shallow and usually alkaline lakes, wet meadows, and sandhills prairie (about 61,000 acres). A bird species checklist is available (233 species, 95 nesters). Upland sandpipers and long-billed curlews are common nesters, as are sharp-tailed grouse and greater prairie-chickens. Burrowing owls are occasional nesters, but any burrows that are large enough for them tend to collapse in the sandy substrate. Several large ranches owned by Ted Turner in the Nebraska Sandhills (McGinley and Dan Hill ranches, in Sheridan County; the Deer Creek, Spikebox, U Cross, and Wright ranches, in Cherry County; and the Blue Creek Ranch, in Garden County) collectively encompass about 316,000 acres. The ranches are used for bison rearing (about nine thousand head) as well as for related wildlife conservation activities, including prairie dog restoration.

NEW MEXICO

Bitter Lake National Wildlife Refuge. 23,000 acres. Federally owned refuge. Manager's address: Box 7, Roswell NM 88201. Shortgrass prairies, shrub-steppe uplands, alkaline marshes, sand dunes, and river bottomlands. A bird species checklist (282 species) is available. Burrowing owls are uncommon breeders.

Bosque del Apache National Wildlife Refuge. 57,191 acres. Federally owned refuge. Manager's address: Box 1246, Socorro NM 87814. Arid grasslands, woody scrub, and wetlands. A bird species checklist (252 species, 95 nesters) is available. Burrowing owls are uncommon nesters. Located in the Arizona–New Mexico Mountains ecoregion.

Bureau of Land Management (BLM). The BLM controls 12.95 million of acres of land in New Mexico (16.6 percent of the total land area), much of which is arid upland habitat that supports shortgrass or scrub-steppe species, including two species of prairie dogs (Gunnison's and black-tailed). Most of these regions are beyond the geographic limits of this book. Most of the black-tailed prairie dog's historic range in southern and southeastern New Mexico consists of BLM lands, which are yet to be completely surveyed for prairie dogs, burrowing owls, and associated grassland species. After more than half a century of intensive and almost unrelenting control efforts, the BLM reported in the late 1990s that only one large black-tailed prairie dog colony (4 to 5 square miles) was still known to occur on BLM land, and all the other known colonies were less than 25 acres. Between 1870 and 1998 the New Mexico population of prairie dogs dropped from an estimated 12 million acres to about 60,000 acres, a 99.5 percent reduction. Surveys in 1996 and 1997 by the BLM in eastern and northeastern New Mexico found only about 140 colonies, far fewer than had been estimated earlier. One important BLM grassland site is the Mescalero Sands Outstanding Natural Area (Chaves County), 6,173 acres of shortgrass and sandsage prairies, supporting one of New Mexico's few remaining flocks of lesser prairie-chickens. This is part of the much larger (645,800-acre) Caprock Wildlife Habitat Area of shortgrass prairie, of which 260,500 acres are publicly owned. For information on specific New Mexico BLM holdings, contact the Bureau of Land Management, J. P. Montoya Federal Building, S. Federal Place, P.O. Box 1449, Sante Fe NM 87404. Prairie dogs (and burrowing owls) are now only very local in New Mexico but have recently reported at the Estancia Valley, 40 miles east of Albuquerque, and in playas on a newly established 2,800-acre nature preserve (La Semilla Nature Park) east of Mesa del Sol (a new subdivision of south Albuquerque). Mesa del Sol is an Environmental Protection Agency development site of about 12,000 acres that will focus on environmental preservation and recycling. The largest known active black-tailed prairie dog site in North America occurs fairly close to the New Mexico–Mexico border near Janos, in northern Chihuahua, Mexico. Information on state-owned natural areas and preserves may be obtained from the New Mexico Department of Game & Fish, P.O. Box 25112, Sante Fe NM 87504.

Capulin Mountain National Monument. 793 acres. Federally owned site. Manager's address: Capulin NM 88414. Upland shortgrass plains and scrubby woodland around an extinct cinder cone volcano that last erupted four to ten thousand years ago and now rises about 1,000 feet above the surrounding landscape. About a hundred bird species have been reported from the monument.

The Gray Ranch. 321,000 acres. A Nature Conservancy site in Hidalgo County, in the Animas Mountain region of southwestern New Mexico's "bootheel," 200 miles east of Tucson. A great diversity of community types, including shortgrass prairie. Grassland species include pronghorns, the rare and highly local white-sided jackrabbit, and the endangered aplomado falcon. Black-tailed prairie dogs are being reestablished here. For information contact The Nature Conservancy, 212 E. Marcy St., Suite 200, Santa Fe NM 87501–2049. No species checklists are available.

Grulla National Wildlife Refuge. 3,236 acres. Federally owned refuge managed through the Muleshoe National Wildlife Refuge (see entry in the Texas section of this appendix). A playa lake surrounded by alkaline grasslands (about 1,000 acres) in Roosevelt County near Portales, along the New Mexico–Texas border. A bird species checklist is available (86 species, 21 nesters). An important wintering area for waterfowl, shorebirds, and sandhill cranes (*Grulla* is the Spanish word for crane). Burrowing owls are common nesters.

Kiowa National Grassland. 136,505 acres. Federally owned site managed by the U.S. Forest Service, 714 Main St., Clayton NM 88415. Shortgrass uplands and the 900-foot-deep Canadian River Canyon. Reaches the Oklahoma border, where it is contiguous with the Rita Blanca National Grassland (see entry below). A bird species checklist (226 species, 60 probable nesters) is available. There are prairie dog colonies (about 600 acres in 1998) and burrowing owls. Thirteen of 18 prairie dog towns surveyed in 1998 also had burrowing owls.

Las Vegas National Wildlife Refuge. 8,750 acres. Federally owned refuge. Manager's address: Rte. 1, Box 399, Las Vegas NM 87701. Native prairie, ponderosa pine and juniper canyons, and wetlands, located southeast of Las Vegas, San Miguel County, at the edge of the Southern Rocky Mountains ecosystem. A bird species checklist (238 species, 76 nesters) is available. Burrowing owls are common breeders; prairie falcons, upland sandpipers, and long-billed curlews also nest; and ferruginous hawks and golden eagles occur in winter.

Maxwell National Wildlife Refuge. 3,584 acres. Federally owned refuge. Manager's address: Box 276, Maxwell NM 87728. Shortgrass prairies, lakes, and agricultural lands. A bird species checklist (185 species, 47 nesters) is available. Prairie dogs are present, and burrowing owls are occasional breeders. Prairie falcons and golden eagles are residents, and ferruginous hawks occur in winter.

Vermejo Park Ranch. 578,000 acres. A Ted Turner bison ranch located north of Cimarron, Colfax County, supporting about 1,400 bison and associated shortgrass wildlife, including about 400 acres of prairie dogs in 2002. Another Ted Turner bison ranch in New Mexico is the Ladder Ranch (300,000 acres, 156,000 deeded), near Caballo, Sierra County, which supports about 900 bison. A third Turner bison ranch is the Armendaris Ranch (358,643 acres), near Engle, Sierra County, holding about 1,200 bison. Black-tailed prairie dogs are being restored at all three ranches, and burrowing owls have been reported at Ladder Ranch and Armendaris Ranch.

NORTH DAKOTA

Alkali Lake. 320 acres. A federally owned site managed by the Bureau of Land Management, 2933 Third Ave. W., Dickinson ND 58601. Located about 31 miles north and a mile east of Williston, Williams County, off County Road 9. An alkaline wetland and native surrounding prairie. Many mixed-grass species are here, including marbled godwit, Baird's sparrow, upland sandpiper, and such shorebirds as piping plover and Wilson's phalarope.

Arrowwood National Wildlife Refuge. 15,934 acres. Federally owned refuge. Manager's address: 7745 11th St. SE, Pingree ND 58476. Located 5 miles north of Edmunds, Stutsman County,. Mixed-grass prairies (8,000 acres of native grasslands) and wetlands in the glaciated James River Valley. A bird checklist (246 species, 105 nesters) is available. Many mixed-grass birds, including Baird's sparrow, marbled godwit, upland sandpiper, sharp-tailed grouse, Swainson's hawk, and northern harrier. A small colony of prairie dogs is present; burrowing owls are rare.

Audubon National Wildlife Refuge. 14,735 acres. Federally owned refuge. Manager's address: R.R. 1, Coleharbor ND 58531. Upland shortgrass and mixed-grass prairies (about 3,000 acres of planted native grasses) and marshes adjoining a large subimpoundment of Lake Sakakawea. A vertebrate checklist (239 bird species, 85 nesters; including Baird's and LeConte's sparrows) is available. Burrowing owls are rare breeders; sharp-tailed grouse and marbled godwits are common. The 37 mammal species include pronghorns, white-tailed jackrabbits, and three ground squirrels, plus 5 reptiles, 3 amphibians, and 37 fish.

Big Gumbo. 20,000 acres. A federally owned area managed by the Bureau of Land Management, 2933 Third Ave. W., Dickinson ND 58601. Includes Cedar Ridge, a 1,900-acre area of shortgrass and sagebrush prairie at the northeastern limits of woody sagebrush in the United States. Located near the northwestern South Dakota and southwestern North Dakota borders in Bowman County, south of Marmarth. The area extends along the West River Road and Camp Crook Road, and follows the Little Missouri River Valley. Arid high plains, with pronghorns, lark buntings, chestnut-collared longspurs, long-billed curlews, ferruginous hawks, and golden eagles. Sage-dependent species, such as greater sage-grouse and Brewer's sparrow, are also present. Undeveloped, with sometimes impassable side roads.

Cedar River National Grassland. 6,237 acres. Federally owned site managed by the U.S. Forest Service, Box 390, Lemmon SD 57638. Shortgrass and mixed-grass prairie along the South Dakota line. Located near Grand River National Grassland (see entry below) and the Standing Rock Indian Reservation (see entry in the North Dakota section of this appendix). Still mostly inaccessible by car. No species checklist is available.

Cross Ranch Nature Preserve. 6,000 acres. A Nature Conservancy preserve southwest of Washburn, McLean County, containing mainly mixed-grass prairie and riparian woods. A herd of about a hundred bison is present. For information contact The Nature Conservancy, 1256 Parkview Dr., Bismarck ND 58501. Nearby is the state-owned Cross Ranch State Park (589 acres). For information contact HC2, Box 152, Sanger ND 58567. A total of 147 species (with at least 33 nesters) have collectively been reported for the preserve and the state park. The burrowing owl is absent, but other High Plains species, such as sharp-tailed grouse and upland sandpipers, are common. There is a preliminary plant list of more than 250 species. Four miles south of Cross Ranch State Park is Smith Grove Wildlife Management Area, with old-growth riparian woods, prairie, and woody draws.

Des Lacs National Wildlife Refuge. 19,554 acres. Federally owned refuge. Manager's address: Box 578, Kenmare ND 58746. Mixed-grass prairie (about 7,000 to 10,000 acres), brushland, marshes, and several small river-dependent lakes. A bird species checklist (a collective list including the J. Clark Salyer and Upper Souris national wildlife refuges; see entries below) is available (308 species, 169 nesters), which include 5 species of nesting grebes and 16 nesting sparrows, longspurs, and towhees. Burrowing owls are only occasional nesters in the Souris Valley.

Fort Berthold Indian Reservation. 980,000 acres. Includes the Mandan, Hidatsa, and Arikara tribes, with about 4,800 reservation residents as of 2000, out of 8,700 total tribal members. No information is available on native fauna of the extensive areas of prairie. In 2003 the total tribal acreage suitable for prairie dogs in North Dakota was estimated at 20,500 acres (Tim Vosburgh, Tribal Black-tailed Prairie Dog Coordinator, personal communication via Robert Luce).

Fort Stevenson State Park. 600 acres. State owned site managed by North Dakota Parks and Tourism, 604 E. Boulevard Ave., Bismarck ND 58501. Located 3 miles south of Garrison, McLean County. This peninsula along the shore of Lake Sakakawea consists mostly of rolling hills and upland prairie, including a prairie dog town. Coyotes, jackrabbits, sharp-tailed grouse, and other grassland animals are present. A few miles south of the east end of Garrison Dam and between the dam spillway and the river is the Riverdale Wildlife Management Area (2,197 acres), managed by the North Dakota Game & Fish Department, Bismarck ND 58501. It has a sharp-tailed grouse display ground and a bird checklist of more than one hundred species.

J. Clark Salyer National Wildlife Refuge. 58,700 acres. Federally owned refuge. Manager's address: Box 66, Upham ND 58789. Located 3 miles north of Upham, North Dakota. A transitional mixed-grass and tallgrass refuge (about 18,000 acres of native prairie) in the Souris River Valley. A bird species checklist (collective for several of the Souris River Valley refuges) is available (266 species, 147 nesters). All of the typical mixed-grass sparrows breed in this largest of North Dakota's national wildlife refuges, including LeConte's, Baird's, and Nelson's sharp-tailed sparrows, plus the other usual northern prairie and prairie wetland species.

Knife River Indian Villages National Historic Site. 1,758 acres. Federally owned historic site managed by the National Park Service, Stanton ND 58571. Located on State Highway 31, half a mile north of Stanton, Mercer County. Grasslands and riverbottom forests plus a restored Mandan earth lodge. A major Mandan and Hidatsa village was here during the Lewis and Clark expedition. A bird checklist (212 species) is available. Locally occurring grassland birds include sharp-tailed grouse, black-billed magpie, bobolink, and loggerhead shrike.

Lake Ilo National Wildlife Refuge. 4,043 acres. Federally owned refuge. Manager's address: Box 127, Dunn Center ND 58626. Located near Dunn Center and Killdeer, Dunn County. Native mixed-grass prairie (1,200 acres) and wetlands plus planted grassland areas. A bird species checklist (205 species, 83 nesters) is available. Shorebirds, including piping plovers, upland sandpipers, and marbled godwits, all breed commonly; burrowing owls are uncommon breeders.

Lewis and Clark State Park. 490 acres. State-owned site managed by North Dakota Parks and Tourism, 604 E. Boulevard Ave., Bismarck ND 58501. Located 16 miles east and 3 miles south of Williston, Williams County, along one of the northern bays of Lake Sakakawea. This rather remote park has one of the largest intact areas of native mixed-grass prairie of all the North Dakota state parks. No species checklists are available.

Little Missouri National Grassland. 1,027,852 acres. Two separate units, each more than 500,000 acres. Federally owned land managed by the U.S. Forest Service, 161 21st. St. W., Dickinson ND 58601; or HC02, Box 8, Watford City ND 58854. Shortgrass plains and eroded badlands. This is the largest area of federally owned shortgrass prairie in the nation, representing about a quarter of all national grassland acreage. There were about 4,000 acres of prairie dog colonies in 2003, and about a third of these colonies supported burrowing owls. All of the typical shortgrass vertebrates are likely to be found here, but specific lists are still lacking. The list for Theodore Roosevelt National Park (see entry below) probably applies to this area as well. The nearby Fort Berthold Indian Reservation (about 1 million acres) is similar in its biota. In 1994 its estimated prairie dog acreage was 500 acres, and in 2002 it had four colonies of prairie dogs. It also maintains a managed bison herd. For information contact Fort Berthold Reservation, New Town ND 58763. Other private bison herds in North Dakota include Hawks Nest Buffalo, near Carrington, Foster County; Rossow Buffalo Ranch, near Flasher, Morton County; RX Ken Mar Buffalo Ranch, near New Rockford, Eddy County; and Dakota Plains Buffalo, near Pingree, Stutsman County.

Little Missouri State Park. 5,900 acres. State-owned site managed by North Dakota Parks and Tourism, 604 E. Boulevard Ave., Bismarck ND 58501. Located 16 miles north and 3 miles east of Killdeer, Dunn County. Badlands and arid grasslands, with prairie falcons, golden eagles, coyotes, and similar fauna. No species checklists are available.

Long Lake National Wildlife Refuge. 22,310 acres. Federally owned refuge. Manager's address: R.R. 1, Moffit ND 58560. Located about a mile east of Moffit, Burleigh County. Mixed-grass prairies and croplands (500 acres) around a 16,000-acre lake. A bird species checklist (203 species, 78 nesters) is available. Many prairie birds occur here, including 20 species of grassland sparrows, such as the Baird's sparrow, chestnut-collared longspur, lark bunting, and clay-colored sparrow. Willets, piping plovers, marbled godwits, Wilson's phalaropes, and upland sandpipers are also regular nesters. Northern harriers also nest here, as do sharp-tailed grouse.

Lostwood National Wildlife Refuge. 26,900 acres. Federally owned refuge. Manager's address: R.R. 2., Box 98, Kenmore ND 58746. Upland mixed grasses (22,000 acres, about two-thirds being original mixed-grass prairie), aspen groveland, and alkaline lakes and marshes or prairie potholes. Burrowing owls are occasional nesters in the Souris Valley, using the burrows of various rodents. Mixed-grass endemics, such as Sprague's pipit and Baird's sparrow, are common, as are sharp-tailed grouse, marbled godwit, McCown's and chestnut-collared longspurs, and most other grassland birds of the northern High Plains. A vertebrate checklist (226 bird species, 104 nesters) is available, which includes 37 mammals, 3 reptiles, and 4 amphibians. Pronghorns are rare, but 3 ground squirrels occur.

Oahe Wildlife Management Area. 23,500 acres. A federally owned site managed by North Dakota Game and Fish. Located 4 miles south of Bismarck on State Highway 1804. Grasslands and riverine hardwood forests, on both sides of the Missouri River. No species checklists are available.

Standing Rock Indian Reservation. 2,300,000 acres. The largest North Dakota Indian reservation (contact: Fort Yates ND 58538), if the South Dakota acreage is included. It was believed to have about 137 prairie dog colonies on 1,000 acres as of 2002. The 2000 Native American population on the reservation was 3,100, out of a total of 13,000 tribal members.

Theodore Roosevelt National Park. 70,446 acres. Federally owned park managed by the National Park Service, Box 7, Medora ND 58645. Eroded badlands, shortgrass plains (about 50,000 acres), and shrubsteppe. A bird species checklist (170 species, 58 nesters) is available. The south unit of the park (46,128 acres) has elk (about 550 in 2003), bison (about 320 in 2003), mule and white-tailed deer and pronghorns, plus coyotes and golden eagles. In the north unit (24,070 acres) are bison (about 300 in 2003), pronghorns (variable), elk (occasional), and bighorn sheep (15 to 20). Prairie dogs are present in both units (an estimated 1,000 acres and 30 colonies in 2001–02), and burrowing owls are uncommon nesters.

Upper Souris National Wildlife Refuge. 32,000 acres. Federally owned refuge. Manager's address: R.R. 1, Foxholme ND 58738. Mixed grasses (about 19,000 acres of native prairie), deciduous woods, and wetlands. A bird species checklist (including several of the Souris River Valley refuges) is available (266 species, 147 nesters). Burrowing owls are only occasional nesters in the Souris Valley (see the Des Lacs NWR entry for representative species). There is a mammal list of 35 species, including pronghorn, white-tailed jackrabbit, and 2 ground squirrels.

Altus-Lugert Wildlife Management Area. 10,400 acres. State-owned site managed by the Oklahoma Department of Wildlife Conservation, 1801 N. Lincoln Blvd, Oklahoma City OK 73101. Located 2 miles east of Granite, Greer County. Prairie uplands along the shoreline of Lake Altus. No species checklists are available.

Beaver River Wildlife Management Area. 15,600 acres. State-owned site managed by the Oklahoma Department of Wildlife Conservation, 1801 N. Lincoln Blvd, Oklahoma City OK 73101. Located 11 miles west of Beaver, Beaver County. Mostly sandsage grassland, with prairie dogs (two colonies), burrowing owls, swift foxes, and lesser prairie-chickens. Beaver State Park (360 acres) similarly has sandsage and sand dune habitats; it is located 1 mile north of Beaver. No species checklists are available.

Black Kettle National Grassland. 32,729 acres. (30,724 acres in Oklahoma; 3,005 acres in Texas). Federally owned site managed by the U.S. Forest Service, Rte. 1, Box 55B, Cheyenne OK 73628. Scattered holdings in Roger Mills County, near Cheyenne and Reydon. Diverse prairies (both mixed-grass and tallgrass) and upland woods. No species checklists are available.

Black Mesa Preserve. 1,600 acres. Owned by The Nature Conservancy but managed by the Oklahoma Tourism and Recreation Department, P.O. Box 52002, Oklahoma City OK 73152–2002. Shortgrass and juniper woodland preserve. Golden eagles, prairie falcons, and pronghorns are present in this transition zone between the shortgrass prairie and the Rocky Mountains. It is contiguous with Black Mesa State Park (549 acres); for information contact HCR 1, Box 8, Kenton OK 73946. Like Black Mesa Preserve, this park includes juniper woodlands and shortgrass prairies. The park has a large prairie dog town, with burrowing owls present. No species checklists are available for either area, but a checklist for the entire Oklahoma panhandle includes over 290 bird species. Both sites are located in Cimarron County, about 27 miles northwest of Boise City, and southeast of Kenton, Oklahoma, in the extreme northwestern panhandle.

Canton Reservoir and Wildlife Management Area. 16,775 acres. Federally owned site managed by the U.S. Army Corps of Engineers. Located about 2 miles northwest of Canton, Blaine County. Mixed-grass prairies, with bottomland and crosstimber upland forests. A prairie dog town and nature trail are located near Canton Reservoir. No species checklists are available.

Ellis County Wildlife Management Area. 4,800 acres. State-owned site managed by the Oklahoma Department of Wildlife Conservation, 1801 N. Lincoln Blvd., Oklahoma City OK 73101. Located 15 miles southeast of Arnett, Ellis County, on Highway 46. Shinnery oak grasslands, with lesser prairie-chickens and other grassland species, plus wooded bottomlands along a lake. No species checklists are available.

Fort Cobb Wildlife Management Area and Fort Cobb State Park. 8,020 acres and 1,872 acres, respectively. One federally owned site and one state-owned site, jointly managed by the Bureau of Reclamation and the Oklahoma Department of Wildlife Conservation, 1801 N. Lincoln Blvd., Oklahoma City OK 73101. Located 6 miles north of Fort Cobb, Caddo County. Native mixed-grass prairie and blackjack oak woodlands. No species checklists are available.

Fort Supply Wildlife Management Area. 5,418 acres. Federally owned site managed by the U.S. Army Corps of Engineers. Located 1 mile south of Fort Supply, Woodward County. Sandsage grasslands and riparian hardwoods, with a variety of grassland species. No species checklists are available.

Guymon Game Reserve and Sunset Lake. 192 acres. City-owned site at the western edge of Guymon, Texas County. A grassland and small lake with confined bison, elk, and a colony of prairie dogs. No species checklists are available.

Optima National Wildlife Refuge. 4,333 acres. Federally owned refuge. Manager's address: Rte. 1, Box 68, Butler OK 73625. Located 2 miles northwest of Hardesky, Texas County. Shortgrass and mixed-grass prairies, sage-dominated shrubsteppe, and hardwoods in the Beaver River Valley. A bird species checklist (246 species, 106 nesters) is available. Burrowing owls are occasional nesters, and pronghorns also occur. The nearby and associated Optima Wildlife Management Area (about 10,000 acres) is managed jointly by the Army Corps of Engineers and the Oklahoma Department of Wildlife Conservation, 1801 N. Lincoln Blvd., Oklahoma City OK 73101. Not far away (5 miles south and a mile east of Hardesky) is Shultz Wildlife Management Area (340 acres), also managed by the Oklahoma Dept. of Wildlife Conservation. This area consists of shortgrass prairie along a wooded stream and supports golden eagles and ferruginous hawks during winter.

Packsaddle Wildlife Management Area. 7,500 acres. State-owned site managed by the Oklahoma Department of Wildlife Conservation, 1801 N. Lincoln Blvd., Oklahoma City OK 73101. Located 16 miles south of Arnett, Ellis County. Shortgrass, sandsage, and shinnery oak grasslands, along with riparian woodlands. No species checklists are available.

Rita Blanca National Grassland/Wildlife Management Area. 93,013 acres (15,600 acres in Oklahoma; 77,413 acres in Texas). Federally owned site managed by the U.S. Fish & Wildlife Service, Box 38, Texline TX 79807. Located about 17 miles southwest of Boise City, Cimarron County, Oklahoma, on Highway 56. Shortgrass prairie with prairie dogs, ferruginous hawks, golden eagles, and mountain plovers. (See also the Rita Blanca National Grassland entry in the Texas section of this appendix.)

Salt Plains National Wildlife Refuge. 32,008 acres. Federally owned refuge. Manager's address: Rte. 1, Box 76, Jet OK 763749. Located 8 miles north of Jet. Native grasslands, brush, and forest, with large areas of barren salt flats. A bird species checklist (243 species, 98 nesters) is available; a mammal list of 30 species includes coyotes and badgers. Burrowing owls are rare in summer. Great Salt Plains State Park (8,890 acres) is nearby and has similar habitats.

Sandy Sanders Wildlife Management Area. 16,400 acres. State-owned site managed by the Oklahoma Department of Wildlife Conservation, 1801 N. Lincoln Blvd., Oklahoma City OK 73101. Located 10 miles south of Erick, Beckham County. Arid grassland and cedar breaks area. No species checklists are available.

Washita National Wildlife Refuge. 8,200 acres. Federally owned refuge. Manager's address: Rte. 1, Box 68, Butler OK 73625. Located about 7 miles northwest of Butler in Custer County. About 3,000 acres of mixed-grass prairies plus wooded creeks and a reservoir. A bird species checklist (220 species, 67 nesters) is available. Prairie dogs are present, and burrowing owls are rare in summer. The state-owned Foss State Park (1,749 acres) is nearby (13 miles west of Clinton) and has similar habitats, plus the adjoining 8,800-acre Foss Reservoir. For information contact HC 66, Box 111, Foss OK 73647.

Waurika Wildlife Management Area. 21,056 acres. Federally owned site managed by the U.S. Army Corps of Engineers. Located 4 miles south of Corum, southwestern Stephens County. Mixed-grass prairie and hardwoods around Waurika Lake. No species checklists are available.

Wichita Mountains National Wildlife Refuge. 59,020 acres. Federally owned refuge. Manager's address: Rte. 1, Box 448, Indiahoma OK 73552. Located northwest of Lawton, Comanche County. Mixed-grass prairie (about 30,000 acres) and juniper and scrub oak uplands up to 1,400 feet high, the eroded remnants of Paleozoic mountains some three hundred million years old, and with more than a hundred small wetlands. Managed partly for bison (about 600), elk (about 500), and longhorn cattle. A bird species checklist (278 species, 78 nesters) is available. There are also species checklists of 50 mammals, 64 reptiles and amphibians, 36 fish, and 806 plants. Prairie dogs are present (about 150 acres), and burrowing owls are rare residents. Within the city of Lawton there is also a small prairie dog colony in Elmer Thomas Park. Immediately south of the refuge is the federally owned Fort Sill military reservation, with 93,000 acres of mostly mixed-grass prairie, providing a major habitat for some rare shrub-dependent species such as black-capped vireos. Access to the military facilities is restricted, and traffic may be limited to public roads. No species checklists are available. Farther east is the Osage Indian Reservation, near Pawhuska, Osage County. The Nature Conservancy's Tallgrass Prairie Preserve received 300 bison in 1993, with plans to increase the herd eventually to 1,850 animals on about 37,000 acres. As of 2000 there were about 650 bison present.

SASKATCHEWAN

Cypress Hills Interprovincial Park. About 85,000 acres. The mixed-grass prairies of Saskatchewan, Alberta, and the Cypress Hills region (straddling the Alberta-Sakatchewan border) are somewhat more mesic and cold adapted than are those farther south and are known as fescue prairie because of their abundance of fescue grasses. These grasslands merge with the aspen grovelands to the north and the coniferous forests of the Rocky Mountain foothills to the west as well as with the more typical mixed-grass prairies to the south. Cypress Hills Park itself is largely wooded with aspen grovelands but has surrounding grasslands. Provincially owned grasslands surrounding the Cypress Hills also extend south from the Trans-Canada Highway, especially from east of Maple Creek, Saskatchewan, to the U.S. border. North and west of Maple Creek are extensive fescue prairies that include the Bigstick Lakes Prairie Farm Rehabilitation Site (about 19,000 acres). In the Bigstick Lakes site are three large saline lakes (Bigstick, Crane, and Ingebright) plus extensive grasslands. These lakes are near Fox Valley and Tompkins. The ecologically similar Alkali Lake is located along the Montana-Saskatchewan border, near Gladmer. East and West Coteau Lakes are also located close to the border, near Minton. These sites are all classified as Important Bird Areas of Canada. Other designated Important Bird Areas in southern Saskatchewan that include prairie habitats are Barber Lake (near Wiseton), Fife Lake (near Rockglen), the Colgate Area (west of Colgate), Kutawagan Lake (near Nakomis), and the Kindersley-Elma Area (between Smiley and Kindersley). All these sites have variably sized areas of native fescue prairie as well as wetlands. Some native fescue prairie also exists along the South Saskatchewan River, from the Alberta-Saskatchewan border to Lancer, Saskatchewan. Wildlife of this general region also includes some shortgrass and sage-steppe species, such as greater sage-grouse, ferruginous hawk, swift fox, and pronghorn. For Cypress Hills Park information in Saskatchewan contact the superintendent, c/o Box 850, Maple Creek SA SON INO.

Grasslands National Park. 233,000 acres. Located southeast of Val Marie (75 miles north of Malta, Montana) and west of Killdeer, Saskatchewan, in two separate blocks. The West Block can best be reached from Wood Mountain, Saskatchewan, and includes the Frenchman River Valley. This is probably Canada's largest remaining area of protected native grasslands and consists mostly of mixed-grass prairie, with some shortgrass areas. The East Block contains badlands topography (the Killdeer Badlands) and shortgrass prairie. Wildlife includes black-tailed prairie dogs (Canada's only population), swift foxes, pronghorns, burrowing owls, greater sage-grouse, prairie falcons, Sprague's pipit, chestnut-collared longspur, and other grassland biota. A bird species list (177 species, nesting species unreported) is available. For information contact P.O. Box 150, Val Marie SK SON 2TO.

Last Mountain Lake National Wildlife Area. About 38,500 acres. One of Canada's largest remaining areas of mixed-grass prairie (which cover about half of the total acreage). Located north of Regina and jointly managed by the Canadian Wildlife Service and the Province of Saskatchewan. Designated as one of Canada's Important Bird Areas.

Old Man on His Back Prairie and Heritage Conservation Area. 13,100 acres. Shortgrass prairie, perhaps the best such natural area left in Saskatchewan. A Nature Conservancy Canada site; for information contact the Canada Conservation Partnership, 1313 Fifth St. S.E., Suite 320, Minneapolis MN 55414.

Saskatchewan Landing Provincial Park. 13,830 acres. A provincial park located south of Kyle on Highway 4 (contact: Box 179, Stewart Valley SK SON 2PO. Mixed-grass (fescue) prairie as well as pronghorns, burrowing owls, prairie falcons, ferruginous hawks, and other grassland species more typical of shortgrass prairie. No species checklists are available.

St. Dinis National Wildlife Area and Bradford National Wildlife Area. 890 acres and 304 acres, respectively. Grasslands (shortgrass, mixed-grass, and cultivated) and other habitats located southeast (Bradford) and east (St. Denis) of Saskatoon. Farther east of Saskatoon and near the Manitoba border is Prairie National Wildlife Area, with twenty-seven parcels of land scattered over about 7,250 acres, with native or cultivated grasslands and vegetation associated with wetlands. For information contact the Canadian Wildlife Service, Twin Atria Bldg., 2nd Floor, 4999 98th Ave., Edmonton AB T6B 2X3.

Stalwart National Wildlife Area. 3,767 acres. A Canadian Wildlife Service area of mixed-grass (fescue) prairie and wetlands. Located northwest of Regina. For information contact the Canadian Wildlife Service, Twin Atria Bldg., 2nd Floor, 4999 98th Ave., Edmonton AB T6B 2X3.

SOUTH DAKOTA

Badlands National Park. 243,302 acres. Federally owned park managed by the National Park Service, Box 6, Interior SD 57750. Eroded high plains with shortgrass prairie and shrubsteppe. A bird species checklist (208 species) is available. There are prairie dog colonies (5,786 acres in 1993), burrowing owls, and reintroduced black-footed ferrets. Bison (500 to 550 head, stocking rate about 400 acres/head), pronghorns, and bighorn sheep also occur. Releases of swift foxes were made in 2003 and will continue until 2005.

Bad River Ranch. 138,310 acres. A Ted Turner ranch located along the Bad River, about 20 miles west of Pierre. A herd of about 2,500 bison is maintained here, one of the largest privately owned herds in South Dakota. Other High Plains wildlife are being actively managed and conserved here, including about 400 acres of prairie dogs in 2002, as well as reintroduced swift foxes. Twenty swift foxes were released in 2002, with additional releases in 2003. Another large bison ranch in South Dakota is the Triple U Buffalo Ranch, also near Fort Pierre (Stanley County), with 3,500 animals. The Triple Seven (777) Ranch near Hermosa (Custer County) has over 20,000 acres and more than 2,000 bison. Other large bison ranches include Mahoney Buffalo and Cattle Ranch, near Fulton (Hanson County); the Black Hills Bison Company, near Piedmont (Lawrence County); Trails End Bison Ranch near Presho (Lyman County); and the Lazy YX Ranch, near Murdo (Jones County). Bison may also be seen at the Buffalo Run Game Reserve, near Gettysburg, Potter County.

Bear Butte State Park and Fort Meade Recreation Area. About 6,700 acres. Bear Butte State Park is managed by the South Dakota Game, Fish & Parks Department, P.O. Box 688, Sturgis SD 57785. The state park is contiguous with the federally owned recreation area. For information about the latter, contact Bear Butte State Park, 10 Roundup Street, Belle Fourche SD 57717. Bear Butte is an igneous mountain outlier (4,422 feet at the summit) of the Black Hills, long considered sacred by Native Americans. Mostly grasslands. A bird species checklist (189 species) is available for the state park and the associated recreation area. Located in Meade County, 6 miles northeast of Sturgis, at the eastern edge of the Black Hills. Nearby is the Broken Heart Ranch, which raises bison under near-wild conditions, with some on Buffalo Gap National Grassland.

Buffalo Gap National Grassland. 591,727 acres. Federally owned site managed by the U.S. Forest Service, 209 N. River, Hot Springs SD 57747. Shortgrass and mixed-grass prairie. Abundant prairie dogs support the largest known wild population of black-footed ferrets (nearly 300 in 2002). A bird species checklist (230 species) is available. Also available are lists of 58 mammals, 19 reptiles, and 8 amphibians, plus lists of nearly 300 plant species, including 92 grasses, 160 forbs (broadleaved herbs), 22 shrubs, and 11 trees. There are prairie dog colonies (about 5,400 acres in 1998) and burrowing owls. In Wall, South Dakota, is the National Grasslands Visitor Center, 708 Main St., Box 425, Wall SD 57790. This comprehensive visitor center provides information on all the national grasslands. For the contiguous Oglala National Grassland, see the entry in the Nebraska section of this appendix.

Cheyenne River Indian Reservation. 1,400,000 acres. Federal Indian reservation located on the west side of the Missouri River and contiguous to the north with the Standing Rock Indian Reservation (see entry in the North Dakota section of this appendix). The Cheyenne River Reservation had about two thousand bison as of 2000 and is managing for the maintenance of 13,000 prairie dog acres (44,000 acres were present in 2004). Black-footed ferrets were released there in fall 2000. In 2003 the total tribal acreage suitable for prairie dogs in South Dakota was estimated at 160,000 acres (Tim Vosburgh, Tribal Black-tailed Prairie Dog Coordinator, personal communication via Robert Luce).

Crow Creek Indian Reservation. 125,591 acres. Headquarters: Fort Thompson SD 57339. Federal Indian reservation located in central South Dakota, on the east side of the impounded Missouri River. This relatively small reservation is managing for maintaining at least 1,000 acres of prairie dogs (450 acres were present in 1994).

Fort Pierre National Grassland. 115,997 acres. Federally owned site managed by the U.S. Forest Service, Box 417, 124 S. Euclid Ave., Pierre SD 57501. Shortgrass and mixed-grass prairie, with a bird species checklist (unpublished) of more than 200 species, including ferruginous hawks, upland sandpipers, marbled godwits, greater prairie-chickens, and sharp-tailed grouse. Prairie dog colonies (about 720 acres in 1998) and burrowing owls also occur.

Grand River National Grassland. 154,900 acres. Federally owned site managed by the U.S. Forest Service, Box 390, Lemmon SD 57638. Shortgrass and mixed-grass prairies, with some badlands. For information contact the Bureau of Indian Affairs, P.O. Box 325, Eagle Butte SD 57625. Along the North Dakota border and close to Cedar River National Grassland (see entry). Most of the area is inaccessible by car. There are prairie dog colonies (about 1,600 acres in 1998) and burrowing owls. High Plains animals include pronghorns, sharp-tailed grouse, and mule deer. No species checklist is available.

Lacreek National Wildlife Refuge. 16,250 acres. Federally owned refuge. Manager's address: Star Rte. 3, Martin SD 57551. Sandhills grasslands, including about 4,000 acres of mixed-grass and sandhills prairie, and wetlands, in the Little White River Valley. A bird species checklist (213 species, 93 nesters) is available. There is a prairie dog colony (about 15 acres), and the burrowing owl is an occasional breeder.

Lower Brule Indian Reservation. 1,342,601 acres. Federal Indian reservation located in central South Dakota directly to the east of Fort Pierre National Grassland (see entry) and on the west side of the impounded Missouri River. For information contact the Bureau of Indian Affairs, P.O. Box 325, Eagle Butte SD 57625. This large reservation supports a small herd of bison (about 150 head) and is trying to maintain about 2,000 acres of prairie dog colonies (4,500 acres were present in 1994). Species lists are not available, but the wildlife is probably similar to that reported for the Fort Pierre National Grassland (see entry above).

Pine Ridge Indian Reservation. 2,048,000 acres (372,000 of which are tribally owned). Federal Indian reservation located in southwestern South Dakota adjoining Badlands National Park. For information contact the Bureau of Indian Affairs, P.O. Box 325, Eagle Butte SD 57625. This enormous reservation historically supported a very large population of prairie dogs. Its prairie dog population has been controlled by extensive poisoning, reducing its formerly estimated 268,000 acres of prairie dog colonies to about 35,000 acres by 1994 but increasing again to nearly 200,000 acres by 2003. A bison herd is currently being developed. Species lists are not available, but the wildlife is probably similar to that reported for the Fort Pierre National Grassland (see entry).

Rosebud Indian Reservation. About 1 million acres (409,000 of which are tribally owned). Federal Indian reservation location in southwestern South Dakota. This reservation had an estimated 200,000 acres of prairie dogs in 2004. Several Dakota reservations allow fee-based recreational shooting of prairie dogs, and some also are developing resident bison herds through the Intertribal Bison Cooperative (http://www.intertribalbison.org). Collectively, the Great Plains reservations of the Dakotas and Montana may support the largest remaining prairie dog towns and associated biotic communities in the country. Species lists are not available for any of the South Dakota reservations, but their wildlife is probably much the same as that reported for the Fort Pierre National Grassland (see entry).

Samuel H. Ordway Jr. Memorial Prairie Preserve. 7,800 acres. Owned and managed by The Nature Conservancy, 1000 West Ave. N., Sioux Falls SC 57104. Located near Leola, South Dakota. The largest mixed-grass and tallgrass prairie preserve in South Dakota, with over three hundred identified plant species. A bison herd (about a hundred head, with expansion planned) is present. No species checklist is available.

Slim Buttes and Custer National Forest. 162,000 acres. An extensive area of shortgrass prairie and ponderosa pine forest in Harding County, northwestern South Dakota, managed by the U.S. Forest Service, P.O. Box 2556, Billings MT 59103. The wildlife is not yet extensively documented but includes prairie dogs, greater sage-grouse, ferruginous hawks, lark buntings, and other shortgrass and mixed-grass species.

Standing Rock Indian Reservation. 2,300,000 acres. (See the Standing Rock Indian Reservation entry in the North Dakota section of this appendix.)

Wind Cave National Park. 28,056 acres. Federally owned park managed by the National Park Service, Hot Springs SD 57747. Shortgrass prairie (about 23,000 acres) and coniferous forest. A bird species checklist (193 species) is available.

Bison (about 300 head), elk, pronghorns, and bighorns occur. Burrowing owls are uncommon, and the prairie dogs here (1,216 acres in 1994) have been intensively studied (e.g., King 1955; Hoogland 1994). Several colonies are easily visible from the highways. The adjacent state-owned Custer State Park (about 73,000 acres) is immediately to the north and is managed by South Dakota Game, Fish & Parks, 445 East Capitol, Pierre SD 57501. It consists mostly of native grasslands (about 75 percent), which are managed for bison (about 1,500 head), bighorn sheep, deer, and other large mammals. It also supports prairie dogs (125 acres in 1987). The nearby Black Hills National Forest supported a few prairie dog colonies as of 2003.

TEXAS

Big Spring State Park. 370 acres. State-owned site managed by Texas Parks and Wildlife, 4200 Smith School Road, Austin TX 78744. Located at the west end of Big Spring, Howard County. A relatively small area of native vegetation supporting a prairie dog town and burrowing owls. No species checklists are available. Another western Texas city that has long had prairie dogs within its city limits is Lubbock, Lubbock County, with a colony in MacKenzie Park, and others on farmlands just east of the city. Lubbock was cited by the Texas Commission on Environmental Quality in 2002 for endangering the groundwater by allowing effluent from cattle operations to drain into prairie dog burrows. City officials responded with a plan to poison the approximately 15,000 prairie dogs living on the city's lands. As of early 2003 some surviving animals were being translocated. Statewide the prairie dog population has dropped from an estimated 56,833,000 acres in 1870 to 22,500 acres in 1998, according to the National Wildlife Federation – a 99.96 percent population reduction (see Table 1 in Chapter 3).

Black Kettle National Grassland. 3,005 acres (plus much larger Oklahoma acreage). See the entry in the Oklahoma section of this appendix for a description of this national grassland, which straddles the borders of both states.

Buffalo Lake National Wildlife Refuge. 7,664 acres. Federally owned refuge. Manager's address: Box 228, Umbarger TX 79091. Shortgrass plains and alkaline wetlands in Randall County, including a 175-acre area of unplowed prairie designated as a national natural landmark and called the "High Plains Natural Area." Buffalo Lake is a shallow, alkaline, and usually dry playa lake. A bird species checklist (344 species, 51 nesters) is available. Burrowing owls are common nesters, and prairie dogs, ferruginous hawks and golden eagles all occur (the raptors present in winter).

Caprock Canyons State Park. 15,310 acres. State-owned site managed by Texas Parks & Wildlife, Box 402, Quitague TX 79255. Near Quitague, in Briscoe County. Some native arid grasslands are present as well as badlands and scrubby mesquite and juniper thickets. Caprock Canyons has a small managed herd of bison (32 head as of 1993) as well as pronghorns and over 175 reported species of birds. An introduced prairie dog town is present, attracting golden eagles in winter. No species checklists are available.

Copper Breaks State Park. 1,933 acres. State-owned site managed by Texas Parks & Wildlife, 777 Park Road 62, Quanah TX 79252. Located 12 miles south of Quanah, Hardeman County. Mesquite and juniper thickets plus restored shortgrass and mixed-grass prairie, managed in large part for longhorn cattle; it was previously a cattle ranch. No species checklists are available.

Gene Howe Wildlife Management Area. 6,713 acres in two tracts. State-owned site managed by Texas Parks & Wildlife, 15412 FM 2266, Canadian TX 79014. High plains and canyons in the northeastern Texas panhandle. The Murphy Unit (889 acres) is mostly shortgrass and old fields located in Liscomb County, near Glazier. The larger Main Unit (about 5,000 acres) is in Hemphill County 7 miles northwest of Canadian. Prairie dogs and other sandsage-adapted species occur, including burrowing owls, pronghorns, and horned lizards, at least in the Main Unit. Lesser prairie-chickens are perhaps more common locally here than elsewhere in Texas, and no other WMA in Texas has prairie dogs. No species checklists are available.

Lake Arrowhead State Park. 524 acres. State-owned site managed by Texas Parks and Wildlife, 15412 FM 2266, Canadian TX 79014. Located southeast of Wichita Falls, in Clay County. Upland prairie and mesquite scrub around a small lake, with a prairie dog town. No species checklists are available.

Lake Meredith National Recreation Area. 44,951 acres. Federally owned site. Manager's address: Box 1460, Fritch TX 79036. Grasslands around a large impoundment of the Canadian River. Species lists (225 birds, 65 mammals, 38 reptiles, and 11 amphibians) are available. Pronghorns, swift foxes, prairie dogs, and burrowing owls are all reportedly present. Alibates Flint Quarries National Monument (1,333 acres) is nearby (P.O. Box 1438, Fritch TX 79036) and has native grasses and a quarry where flint for making stone implements was obtained by pre-Columbian ("Panhandle Aspect") Native Americans. No species checklists are available.

Lake Rita Blanca State Park. 1,668 acres. State-owned site managed by Texas Parks & Wildlife, 15412 FM 2266, Canadian TX 79014. Located just south of Dalhart, in Hartley County. Shortgrass prairie around a small playa lake.

Lyndon B. Johnson National Grassland. 20,320 acres. Federally owned site managed by the U.S. Forest Service, P.O. Box 507, Decatur TX 76234. Located northwest of Decatur, in Wise County. Mixed-grass prairies with interspersed oak woodlands. No species checklists are available.

Matador Wildlife Management Area. 28,183 acres. State-owned site managed by Texas Parks & Wildlife, 3036 FM 3256, Paducah TX 79248. Located in Cottle County, 8 miles north and 2 miles west of Paducah. Shinnery oak grasslands, mesquite uplands, gravelly hills, and bottomland. Wild turkey, mule deer, coyotes, bobcats, and other grassland and brush-adapted fauna. There are 73 miles of barely improved roads. No species checklists are available.

McClellan Creek National Grassland. 1,449 acres. Federally owned site managed by the U.S. Forest Service through Black Kettle National Grassland, P.O. Box 266, Cheyenne OK 73628 (see the Black Kettle National Grassland entry in the Oklahoma section of this appendix). Located to the west of that grassland in Gray County, near Pampa and Alanreed. Lake McClellan is a reservoir that was impounded around a natural playa lake; many other shallow playa lakes are in the area. Mostly shortgrass prairie. Pronghorns occur but are rare. No species checklists available.

Muleshoe National Wildlife Refuge. 5,809 acres. Federally owned refuge. Manager's address: Box 549, Muleshoe, Texas 79347. Sandy shortgrass and semidesert grasslands in Bailey County, with alkaline playa lakes that are important for wintering cranes and waterfowl. Bird species checklist available (282 species, 59 nesters). Burrowing owls are common summer nesters, golden eagles occur in winter.

Palo Duro Canyon State Park. 16,402 acres. State-owned, and managed by Texas Parks & Wildlife. Manager's address: Rte. 2, Box. 285, Canon, Texas 79015. Located in Randall County, 12 miles east of Canon, Texas. Shortgrass prairie along the canyon rim, plus, mesquite scrub thickets, and an 800-foot-deep river canyon. No species checklists available. Golden eagles are regular in winter and a few remain to nest; mountain lions and mule deer both occur.

Pat Murphy Wildlife Management Area. 889 acres. State-owned, and managed by Texas Parks & Wildlife, 15412 FM 2266, Canadian, Texas 79014. Located in the extreme northeastern panhandle, three miles south of Lipscomb, Lipscomb County. Shortgrass prairie (432 acres), sandsage, and other arid habitats. Lesser prairie-chickens, mule deer, coyotes, rattlesnakes, and other associated wildlife. No species checklists available.

Playa Lakes Wildlife Management Area. 1,492 acres in three units. State-owned, and managed by Texas Parks & Wildlife, Box 1, Rt. 46, Paducah, Texas 79248. Three separate units, including the Taylor Lakes Unit (530 acres, 7 miles southeast of Clarendon, Donley County), with four natural lakes; the Armstrong Unit (144 acres, 12 miles southwest of Dimmitt, Castro County), with one lake;

and the Dimmitt Unit (422 acres, near Dimmitt; no public access). All units have playa lakes, with the primary function of protecting migratory waterfowl and shorebirds. Taylor Lake has an observation blind. The onetime shortgrass grasslands are gradually being restored to native species. No species checklists are available.

Rita Blanca National Grassland/Wildlife Management Area. 93,273 acres. (77,413 acres in Texas; 15,860 acres in adjoining Oklahoma). Federally owned site managed by the U.S. Forest Service, Box 38, Texline TX 79807. Shortgrass and mixed-grass prairies, with about forty species of grasses present. The grasslands also extend into New Mexico and are the largest area of contiguous grasslands in Texas. No bird species checklist is available. There are pronghorns, swift foxes, prairie dogs (about 950 acres in 1998), and burrowing owls. A total of twenty-one of the forty-one prairie dog towns surveyed had burrowing owls in 1998.

WYOMING

Boysen State Park. 39,545 acres. State-owned park managed by the Wyoming State Parks and Historic Sites Department, 2301 Central Ave., Cheyenne WY 82002. A park sited around a reservoir, with about 20,000 land acres, including some grasslands. Located north of Shoshoni,, in Fremont County. No species checklists are available.

Bureau of Land Management (BLM). The BLM controls 17.5 million acres of land in Wyoming (28 percent of the state's total land area), much of which is arid upland habitat that supports shortgrass or scrub-steppe species, including prairie dogs (both white-tailed and black-tailed). After it was thought to have become extinct in the mid-1900s, the black-footed ferret was rediscovered among colonies of white-tailed prairie dogs on BLM land near Meeteetse. For information on specific Wyoming BLM holdings, contact the state BLM headquarters, 2515 Warren Ave., P.O. Box 1828, Cheyenne WY 82003, or the BLM Casper District, 1701 East E. St., Casper WY 82601. One example of the BLM's large sage-steppe holdings in east-central Wyoming (48,400 acres) occurs along the Middle Fork of the Powder River, Johnson County, and is administered by the regional BLM office (P.O. Box 670, Buffalo WY 82834). Greater sage-grouse, golden eagles, and other sage-adapted species are present, as well as species more typical of ponderosa pine forest. No species checklists are available. In the same county and with similar habitats are the state-managed Bud Love Wildlife Management Area (8,000 acres), northwest of Buffalo, and Taylor Wildlife Management Habitat Area (10,158 acres), west of Kacee, both with shortgrass prairie, scrub, and coniferous woods. For information contact the Wyoming Game and Fish Department, Cheyenne WY 82002.

In Sheridan County, a 560-acre BLM Recreation and Public Purpose site had 268 acres of black-tailed prairie dog colonies in 1979, and a coal-lease site (the Wildcat Preference Right Lease Application Site) of 4,500 acres in Campbell County had 1,166 colony acres of prairie dogs. Among all seven prairie dog sites surveyed, 34 associated species of mammals were seen, as well as 62 bird species, 8 reptiles, and 2 amphibians. BLM lands of southwestern and south-central Wyoming in Uinta and Carbon counties (in the Great Basin ecoregion of The Nature Conservancy) totaling about 149,000 acres supported 10,744 acres of white-tailed prairie dog colonies in 1979. Black-footed ferret releases have been attempted in the Shirley Basin, Carbon County.

Devils Tower National Monument. 1,347 acres. Federally owned site managed by the National Park Service. Manager's address: Devils Tower WY 82714. A 1,200-foot column of volcanic basalt surrounded by ponderosa pines, scrubby thickets, and grasslands. A bird species checklist (158 species, 75 nesters) is available. Burrowing owls are not reported, but prairie dogs are present (19 acres in 1994). Keyhole State Park (15,674 acres) is nearby, with 6,256 acres of prairie. It is managed by the Wyoming State Parks and Historic Sites Department, 2301 Central Ave., Cheyenne WY 82002.

Glendo State Park. 22,430 acres. State-owned park managed by the Wyoming State Parks and Historic Sites Department, 2301 Central Ave., Cheyenne WY 82002. The 9,930 acres of land surrounding an impoundment include some grasslands. Located near Glendo, Platte County. No species checklists are available.

Guernsey State Park. 6,227 acres. State-owned park managed by the Wyoming State Parks and Historic Sites Department, 2301 Central Ave., Cheyenne WY 82002. Partly an impoundment, with about 4,000 acres of land, including some prairie and sage-steppe, as well as pines and juniper woodlands. Located north of Guernsey, Platte County. No species checklists are available.

Hutton Lake National Wildlife Refuge. 2,000 acres. Federally owned refuge located at the western edge of the Great Plains and in the Colorado Rocky Mountains ecoregion of The Nature Conservancy. Managed through Arapaho National Wildlife Refuge, Box 457, Walden CO 80480. Located about 12 miles south of Laramie, Albany County. Native sage grasslands (about 1,200 acres), with pronghorns and prairie dogs. A bird species checklist (153 species, 61 nesters; including golden eagles) is available. Burrowing owls are not listed, and the prairie dogs are the white-tailed species. Nearby (12 miles northwest of Laramie) is the smaller Bamforth National Wildlife Refuge (1,166 acres). It mostly consists of alkaline-adapted greasewood shrub-steppe, with only a small amount of grasslands. It is also managed through Colorado's Arapaho NWR. In the same general area of southeastern Wyoming is the state's largest bison ranch,

Terry Bison Ranch. It is located in Laramie County, south of Cheyenne along the Wyoming-Colorado line, and supports about 2,500 bison on 27,000 acres.

Seminoe State Park. 16,970 acres. State-owned park managed by the Wyoming State Parks and Historic Sites Department, 2301 Central Ave., Cheyenne WY 82002. Situated around a reservoir, with about 5,000 acres of uplands. Grassland and scrub are typical habitats. Located 34 miles north of Sinclair, in Carbon County. In the same county are the Morgan Creek Wildlife Habitat Management Area (4,125 acres), 30 miles north of Sinclair, and the Laramie Peak Wildlife Habitat Management Area (11,000 acres), 40 miles west of Wheatland, Platte County. Both sites have extensive grasslands and sage-steppe. They are managed by the Wyoming Game and Fish Department, 5400 Bishop Blvd., Cheyenne WY 82006. Both are in the Great Basin ecoregion of the Nature Conservancy, and any prairie dogs here would probably be the white-tailed species. No species checklists are available.

Table Mountain Wildlife Habitat Management Area. 1,716 acres. BLM-owned site managed by Wyoming Game and Fish Department, 5400 Bishop Blvd., Cheyenne WY 82006. Located a few miles south of Huntley, Goshen County. Shortgrass prairie, in the North Platte Valley near the Nebraska border, with longspurs, ferruginous hawks, and other grassland birds. No species checklists are available.

Thunder Basin National Grassland. 572,211 acres. Federally owned site managed by the U.S. Forest Service, 2550 E. Richards St., Douglas WY 82633. Shortgrass and shrubsteppe plains between 3,600 and 5,200 feet elevation. The largest area of federally protected grasslands in Wyoming, supporting the state's biggest herd of pronghorns plus mule deer and other sage-steppe species. There are prairie dog colonies (about 18,200 acres in 1998, the largest collective colony acreage of any of the national grasslands). However, sylvatic plague in 2001 caused great losses here. Black-footed ferrets may eventually be introduced on this grassland. A bird species checklist (231 species) is available, with the burrowing owl, greater sage-grouse, mountain plover, long-billed curlew, and both longspurs regular breeders. The Douglas office of the Forest Service also has maps of prairie dog colonies in the Thunder Basin area.

Whitney Preserve. 4,600 acres. A Nature Conservancy preserve in the northern Black Hills region, located near the South Dakota border, in Niobrara County, southwest of Hot Springs, South Dakota. There are about 2,000 acres of prairie in the preserve. Entrance by permission only. For information contact The Nature Conservancy, 258 Main St., Suite 200, Lander WY 82520.

Appendix 2

Scientific Names of Animals in the Text

African lion. *Panthera leo*
American avocet. *Recurvirostra americana*
American crow. *Corvus brachyrhynchos*
American kestrel. *Falco sparverius*
Aplomado falcon. *Falco femoralis*
Badger. *Taxidea taxus*
Baird's sparrow. *Ammodramus bairdii*
Bald eagle. *Haliaeetus leucocephalus*
Barn swallow. *Hirundo rustica*
Bears. *Ursus* spp.*
Beaver. *Castor canadensis*
Bighorn (mountain) sheep. *Ovis canadensis*
Bison (buffalo). *Bison bison*
Black-billed magpie. *Pica pica*
Black-capped vireo. *Vireo atricapillus*
Black-footed ferret. *Mustela nigripes*
Black-tailed jackrabbit. *Lepus californicus*
Black-tailed prairie dog. *Cynomys ludovicianus*
Bluebirds. *Sialia* spp.
Blue-winged teal. *Anas discors*
Bobcat. *Felis rufus*
Bobolink. *Dolichonyx oryzivorus*
Brewer's blackbird. *Euphagus cyanocephalus*
Brewer's sparrow. *Spizella breweri*
Brown-headed cowbird. *Molothrus ater*
Bullsnake. *Pituophis melanoleucus sayi*
Buntings. *Passerina* spp.
Burrowing owl. *Athene cunicularia*
Buteo hawk. *Buteo* spp.
Caribou. *Rangifer tarandus*
Cassin's sparrow. *Aimophila cassinii*

Chestnut-collared longspur. *Calcarius ornatus*
Clay-colored sparrow. *Spizella pallida*
Cliff swallow. *Petrochelidon pyrrhonota*
Columbian mammoth. *Mammuthus columbi*
Common nighthawk. *Chordeiles minor*
Common poorwill. *Phalaenoptilus nuttallii*
Cottontails. *Sylvilagus* spp.
Coyote. *Canis latrans*
Crow (= American crow). *Corvus brachyrhynchos*
Dark-eyed junco. *Junco hyemalis*
Deer mouse. *Peromyscus maniculatus*
Desert cottontail. *Sylvilagus audubonii*
Diamondback rattlesnake. *Crotalis atrox*
Domestic cow. *Bos taurus*
Domestic goat. *Capra hircus*
Domestic horse. *Equis caballos*
Domestic sheep. *Ovis aries*
Eastern cottontail. *Sylvilagus floridanus*
Eastern kingbird. *Tyrannus tyrannus*
Eastern meadowlark. *Sturnella magna*
Eastern yellow-bellied racer. *Coluber constrictor flaviventris*
Elk (wapiti). *Cervus elaphus*
Eskimo curlew. *Numenius borealis*
European starling. *Sturnus vulgaris*
Ferruginous hawk. *Buteo regalis*
Foxes. *Vulpes* and *Urocyon* spp.
Frogs. Ranidae
Golden eagle. *Aquila chrysaetos*
Gopher snake. *Pituophis melanoleucus*
Grasshopper mice. *Onychomys* spp.
Grasshopper sparrow. *Ammodramus savannarum*
Gray fox. *Urocyon cineroargenteus*
Gray (prairie) wolf. *Canis lupus*
Great horned owl. *Bubo virginianus*
Great Plains narrowmouth toad. *Gastrophryne olivacea*
Great Plains skink. *Eumeces obsoletus*
Great Plains toad. *Bufo cognatus*
Greater sage-grouse. *Centrocercus urophasianus*
Green-winged teal. *Anas crecca*
Grizzly bear. *Ursus arctos*

Grosbeaks. *Pheucticus* spp.
Ground squirrels. *Spermophilus* spp.
Gunnison's prairie dog. *Cynomys gunnisoni*
Gyrfalcon. *Falco rusticolus*
Hispid pocket mouse. *Perognathus hispidus*
Horned lark. *Eremophila alpestris*
Ivory-billed woodpecker. *Campephilus principalis*
Kangaroo rats. *Dipodomys* spp.
Killdeer. *Charadrius vociferus*
Kit fox. *Vulpes velox macrotis*
Lark bunting. *Calamospiza melanocorys*
Lark sparrow. *Chondestes grammacus*
LeConte's sparrow. *Ammodramus leconteii*
Lesser earless lizard. *Holbrookia maculata*
Lesser prairie-chicken. *Tympanuchus pallidicinctus*
Lizards. Iguanidae
Loggerhead shrike. *Lanius ludovicianus*
Long-billed curlew. *Numenius americanus*
Long-tailed weasel. *Mustela frenata*
Magpie (= black-billed magpie). *Pica pica*
Mallard. *Anas platyrhynchos*
Mammoth. *Mammuthus* spp.
Marbled godwit. *Limosa fedoa*
Marmots. *Marmota* spp.
Mastodon. Mammutidae
McCown's longspur. *Calcarius mccownii*
Meadowlarks. *Sturnella* spp.
Merlin. *Falco columbarius*
Mexican prairie dog. *Cynomys mexicanus*
Mice. Cricetidae
Mountain lion. *Puma concolor*
Mountain plover. *Charadrius montanus*
Mourning dove. *Zenaida macroura*
Mule deer. *Odocoileus hemionus*
Muskox. *Ovibos moschatus*
Nelson's sharp-tailed sparrow. *Ammodramus nelsoni*
Nine-banded armadillo. *Dasypus novemcinctus*
Northern flicker. *Colaptes auratus*
Northern goshawk. *Accipiter gentilis*
Northern grasshopper mouse. *Onychomys leucogaster*

Northern harrier. *Circus cyaneus*
Northern pintail. *Anas acuta*
Ord's kangaroo rat. *Dipodomys ordii*
Orioles. *Icterus* spp.
Ornate box turtle. *Terrapena ornata*
Peregrine falcon. *Falco peregrinus*
Pheasant (= ring-necked pheasant). *Phasianus colchicus*
Piping plover. *Charadrius melodus*
Plains harvest mouse. *Reithrodontomys montanus*
Plains pocket gopher. *Geomys bursarius*
Pocket gophers. *Thomomys* spp.
Prairie-chickens. *Tympanuchus* spp.
Prairie falcon. *Falco mexicanus*
Prairie kingsnake. *Lampropeltis getulus nigritus*
Prairie rattlesnake. *Crotalus viridis*
Prairie vole. *Microtus ochrogaster*
Pronghorn. *Antilocapra americana*
Raccoon. *Procyon lotor*
Rattlesnakes. *Crotalis* spp.
Red fox. *Vulpes vulpes*
Red-tailed hawk. *Buteo jamaicensis*
Red-winged blackbird. *Agelaius phoeniceus*
Rhinoceros. Rhinocerotidae
Richardson's ground squirrel. *Spermophilus richardsoni*
Rock pigeon (= rock dove). *Columba livia*
Rock wren. *Salpinctes obsoletus*
Rough-legged hawk. *Buteo lagopus*
Sagebrush lizard. *Sceloperus graciosus*
Screech-owls. *Otus* spp.
Sharp-tailed grouse. *Tympanuchus phasianellus*
Short-eared owl. *Asio flammeus*
Skunks. *Mephitis, Conepatus* and *Spilogale*
Snowy owl. *Nyctea scandiaca*
Southern plains woodrat. *Neotoma micropus*
Spotted ground squirrel. *Spermophilus spilosoma*
Sprague's pipit. *Anthus spragueii*
Striped skunk. *Mephitis mephitis*
Swainson's hawk. *Buteo swainsoni*
Swift fox. *Vulpes velox velox*
Texas horned lizard. *Phrynosoma cornutum*

Thirteen-lined ground squirrel. *Spermophilus tridecemlineatus*
Tiger salamander. *Ambystoma tigrinum*
Toads. Bufonidae
Tortoises. *Gopherus* spp.
Towhees. *Pipilo* spp.
Turkey vulture. *Cathartes aura*
Turtles. Chelonia
Upland sandpiper. *Bartramia longicauda*
Utah prairie dog. *Cynomys parvidens*
Vesper sparrow. *Pooecetes gramineus*
Violet-green swallow. *Tachycineta thalassina*
Voles. *Microtus* spp.
Weasels. *Mustela* spp.
Western diamondback rattlesnake. *Crotalus atrox*
Western kingbird. *Tyrannus verticalis*
Western meadowlark. *Sturnella neglecta*
White-footed mouse. *Peromyscus leucopus*
White-sided jackrabbit. *Lepus callotis*
White-tailed deer. *Odocoileus virginianus*
White-tailed jackrabbit. *Lepus townsendii*
White-tailed prairie dog. *Cynomys leucurus*
White-throated swift. *Aeronautes saxatalis*
Whooping crane. *Grus americana*
Wildebeests. *Connochaetes* spp.
Willet. *Catoptrophorus semipalmatus*
Wilson's phalarope. *Phalaropus tricolor*
Wolf (= gray wolf). *Canis lupus*
Woodchuck. *Marmota monax*
Wood-pewees. *Contopus* spp.
Wyoming pocket mouse. *Perognathus fasciatus*
Yellow rail. *Coturnicops noveboracensis*
Yellow-headed blackbird. *Xanthocephalus xanthocephalus*

* The abbreviation "spp." means two or more species.

References

1. THE WESTERN SHORTGRASS PRAIRIE

Allen, D. 1967. *The Life of Prairies and Plains*. New York: McGraw-Hill.

Axelrod, D. I. 1985. Rise of the grassland biome, central North America. *Botanical Review* 51:163–201.

Bachand, R. R. 2001. *The American Prairie: Going, Going, Gone?* Boulder CO: National Wildlife Federation.

Bailey, R. G. 1995. *Description of the Ecoregions of the United States*. Forest Service Miscellaneous Publication No. 1391 (revised). Washington DC: U.S. Forest Service.

Brown, J. E. 1992. *Animals of the Soul: Sacred Animals of the Oglala Sioux*. Rockport MA: Element.

Burroughs, R. D. 1961. *The Natural History of the Lewis and Clark Expedition*. East Lansing: Michigan State University Press.

Dort, W., Jr., and J. K. Jones, eds. 1970. *Pleistocene and Recent Environments of the Great Plains*. Lawrence: University Press of Kansas.

Fishbein, S. 1989. *Yellowstone Country: The Enduring Wonder*. Washington DC: National Geographic Society.

Frazier, I. 1989. *Great Plains*. New York: Farrar Straus Giroux.

Graham, A. 1999. *Late Cretaceous and Cenozoic History of North American Vegetation North of Mexico*. New York: Oxford University Press.

Hart, R. H., and J. A. Hart. 1997. Rangelands of the Great Plains before European settlement. *Rangelands* 19:4–11.

Jones, S. R., and R. C. Cushman. 2004. *A Field Guide to the North American Prairie*. Boston: Houghton Mifflin.

Knight, D. 1994. *Mountains and Plains: The Ecology of Wyoming Landscapes*. New Haven CT: Yale University Press.

Leopold, E. B, and M. F. Denton. 1987. Comparative age of grassland and steppe east and west of the northern Rockies. *Annals of the Missouri Botanical Garden* 74:841–67.

Lynch, W. 1984. *Married to the Wind: A Study of the Prairie Grasslands*. North Vancouver BC: Whitecap Books.

McClaren, M. P., and T. R. Van Devender, eds. 1995. *The Desert Grassland*. Tucson: University of Arizona Press.

Maximilian, Prince of Wied. 1906. *Travels in the Interior of North America, 1832–1834*.

English translation, *Early Western Travels*, ed. R. G. Twaites, vols. 22–24 (Cleveland: Arthur H. Clark, 1904–1907).

McGregor, R. L., and T. M. Barkley, eds. (Great Plains Flora Association). 1986. *Flora of the Great Plains*. Lawrence: University Press of Kansas.

Mengel, R. M. 1970. The North American central plains as an isolating agent in bird speciation. In *Pleistocene and Recent Environments of the Central Great Plains*, ed. W. Dort and J. K. Jones, 280–340. Lawrence: University Press of Kansas.

The Nature Conservancy. 1999. *Ecoregional Map of the United States*. Arlington VA: The Nature Conservancy.

North American Bird Conservation Initiative Committee (NABCIC). 2000. *Bird Conservation Region Descriptions. A Supplement to the North American Bird Conservation Initiative Bird Conservation Regions Map (and Associated Map)*. Arlington VA: NABCIC.

Omernik, J. M. 1987. Ecoregions of the coterminous United States. *Annals of the Association of American Geographers* 77:118–25.

Raventon, E. 1994. *Island in the Plains: A Black Hills Natural History*. Boulder CO: Johnson Printing.

Risser, P. G., E. C. Birney, H. D. Bloeker, S. W. May, W. J. Parton, and J. A. Weins. 1981. *The True Prairie Ecosystem*. Stroudsburg PA: Hutchinson Russ.

Rockwell, D. 1998. *The Nature of North America*. New York: Berkley Books.

Schultz, C. B. 1934. The Pleistocene mammals of Nebraska. *Nebraska State Museum Bulletin* 1:357–92.

Sears, P. B. 1969. *Lands beyond the Forest*. Englewood Cliffs NJ: Prentice-Hall.

Sims, P. L. 1988. Grasslands. In *North American Terrestrial Vegetation*, ed. M. G. Barbour and W. D. Billings, 256–86. Cambridge: Cambridge University Press.

Thornbury, W. D. 1965. *Regional Geomorphology of the United States*. New York: Wiley.

Trimble, D. E. 1990. *The Geologic Story of the Great Plains*. Medora ND: Nature and History Association. (Reprint of *U.S. Geological Survey Bulletin* 1493 [1980].)

U.S. Department of the Interior. 1970. *The National Atlas*. Washington DC: U.S. Geological Survey.

Vankat, J. L. 1979. *The Natural Vegetation of North America*. New York: Wiley.

Wayne, W. J., et al. 1991. Quaternary geology of the northern Great Plains. In *The Geology of North America*, ed. R. B. Morrison, vol. K-2, *Quaternary Non-glacial Geology*, 441–76. Boulder CO: Geological Society of America.

Webb, W. P. 1931. *The Great Plains*. New York: Grosset & Dunlap.

Wells, P. V. 1970. Postglacial vegetational history of the Great Plains. *Science* 167:1574–82.

West, N. E. 1988. Intermountain deserts, shrub steppes, and woodlands. In *North American Terrestrial Vegetation*, ed. M. G. Barbour and W. D. Billings, 209–30. Cambridge: Cambridge University Press.

Williams, T. T. 2001. *Red: Passion and Patience in the Desert*. New York: Pantheon Books.

Winckler, S. 2004. *Prairie: A North American Guide*. Iowa City: University of Iowa Press.

2. A BUFFALO NATION

For literature surveys of bison, see Lott (2002), Geist (1996), Plumb et al. (1992), McHugh (1972), and Roe (1970).

Albrecht, S. 2000. Bison population of the world. *Bison World* 25(3): 41–42.

Allen, J. A. 1876. The American bison, living and extinct. *Memoirs of the Museum of Comparative Zoology at Harvard University* 4(10): 1–246.

Anonymous. 2000. Why bison? *Bison World* 25(1): 46.

Branch, E. D. 1929. *The Hunting of the Buffalo*. New York: Appleton.

Bowyer, R. T., X. Manteca, and A. Hoymorck. 1998. Scent marking in American bison: Morphological and spatial characteristics of wallows and rubbed trees. In *International Symposium on Bison Ecology and Management in North America*, ed. L. Irby and J. Knight, 81–91. Bozeman: Montana State University Press.

Callenbach, K. 2000. *Bring Back the Buffalo! A Sustainable Future for America's Great Plains*. Berkeley: University of California Press.

Cid, M. S. 1987. Prairie dog and bison grazing effects on maintenance of attributes of prairie dog colony. PhD diss., Colorado State University. Abstract in *Dissertation Abstracts International* B Sci. Eng. 49:998.

———, J. K. Detling, A. D. Whicker, and M. A. Brizuela. 1991. Vegetational responses of a mixed-grass prairie site following exclusion of prairie dogs and bison. *Journal of Range Management* 44:100–105.

Conley, K., and S. Albrecht. 2000. History of the bison industry. *Bison World* 25(1): 21–22.

Coppock, D. L., J. K. Detling, J. L. Dodd, and M. I. Dyer. 1980. Bison–prairie dog–plant interactions in Wind Cave National Park, South Dakota. *Proceedings of Conference on Science in the National Parks* 12:184.

———, J. E. Ellis, J. K. Detling, and M. I. Dyer. 1983. Plant-herbivore interactions in a North American mixed-grass prairie. II. Responses of bison to modification of vegetation by prairie dogs. *Oecologia* 56:10–15.

Danz, H. P. 1997. *Of Bison and Man*. Niwot: University Press of Colorado.

Dary, D. A. 1989. *The Buffalo Book: The Full Saga of the American Animal*. Chicago: Swallow Press/Ohio University Press.

Epp, H., and I. Dyck. 2002. Early human-bison population interdependence in the plains ecosystem. *Great Plains Research* 12:323–37.

Fitzgerald, D. 1998. *Bison: Monarch of the Plains*. Portland OR: Graphic Arts Center.

Gard, W. 1960. *The Great Buffalo Hunt*. New York: Knopf.

Geist, V. 1996. *Buffalo Nation: A History and Legend of the North American Bison*. Stillwater MN: Voyageur.

Godreau, V., G. Bornette, B. Oertli, F. Chambaud, D. Oberti, and E. Craney. 1998. Mammalian herbivores: Ecosystem-level effects in two national parks. *Wildlife Society Bulletin* 26:438–48.

Haines, F. 1995. *The Buffalo: The Story of American Bison and Their Hunters from Prehistoric Times to the Present.* Rev. ed. Norman: University of Oklahoma Press.

Irby, L. R., and J. E. Knight, eds. 1998. *Bison Ecology and Management in North America.* Symposium Proceedings, Montana State University, Bozeman.

Isenberg, A. C. 2000. *The Destruction of the Bison: An Environmental History.* Cambridge MA: University of Cambridge Press.

Jacobs, L. 1991. *Waste of the West: Public Lands Ranching.* Tucson: Free Our Public Lands.

Josephy, A. M., Jr. 1982. *Now That the Buffalo's Gone.* New York: Knopf.

Knapp. A. K., J. M. Blair, J. M. Briggs, S. L. Collins, D. C. Hartnett, L. C. Johnson, and E. G. Towne. 1999. The keystone role of bison in North American tallgrass prairie. *BioScience* 49:39–50.

Knowles, C. J. 1986. Some relationships of black-tailed prairie dogs to livestock grazing. *Great Basin Naturalist* 46:198–203.

Krueger, K. A. 1984. An experimental analysis of interspecific feeding relationships among bison, pronghorn and prairie dog. Abstract. *Bulletin of the Ecological Society of America* 65:267.

———. 1986a. Feeding relationships among bison, pronghorn and prairie dogs: An experimental analysis. *Ecology* 67:760–70.

———. 1986b. Interactions and activity patterns of bison and prairie dogs at Wind Cave National Park: Implications for managers. General Technical Report INT-212, pp. 203–8. Ogden UT: U.S. Forest Service.

Larson, F. 1940. The role of bison in maintaining the short grass plains. *Ecology* 21:113–21.

Linderman, F. B. 1962. *Plenty-coups, Chief of the Crows.* Reprint, Lincoln: University of Nebraska Press, 2002.

Lott, D. F. 1981. Sexual behavior and intersexual strategies in American bison. *Zietschrift für Tierpsychologie* 56:97–114.

———. 2002. *American Bison: A Natural History.* Berkeley: University of California Press.

McHugh, T. 1958. Social behavior of the American buffalo (*Bison bison bison*). *Zoologica* 43:1–40.

———. 1972. *The Time of the Buffalo.* New York: Knopf.

O'Brien, D. 2002. *Buffalo for the Broken Heart: Restoring Life to a Black Hills Ranch.* New York: Random House.

Peden, D. G., G. M. Van Dyne, R. W. Rice, and R. M. Hansen. 1974. The trophic ecology of *Bison bison* L. on shortgrass plains. *Journal of Applied Ecology* 11:489–98.

Plumb, G. E., J. L. Dodd, and J. B. Stelfox. 1992. A bibliography on bison. Agricultural Extension Station, University of Wyoming, Laramie.

Reinhardt. V. 1985. Social behavior in a confined buffalo herd. *Behaviour* 92:209–26.

Roe, F. G. 1970. *The North American Buffalo: A Critical Study of the Species in Its Wild State.* 2nd ed. Toronto: University of Toronto Press.

Runestad, T., and S. Albrecht. 2000. Where the buffalo roam: Public herds. *Bison World* 25(1): 76–78.

———. 2000. Bison from nose to tail. *Bison World* 25(1): 162–65.

Schwartz, C. C., and J. E. Ellis. 1981. Feeding ecology and niche separation in some native and domestic ungulates on the shortgrass prairie. *Journal of Applied Ecology* 18:343–53.

Seton, E. T. 1909. *Lives of Northern Mammals: An Account of the Mammals of Manitoba.* New York: Scribner's.

Shaw, J. H. 1998. Bison ecology – what we do and do not know. In *Bison Ecology and Management in North America, Symposium Proceedings,* ed. L. I. Irby and J. E. Knight, 113–20. Bozeman: Montana State University.

———. 2000. How many bison originally populated western rangelands? *Bison World* 25(1): 38–41.

Sims, P. L. 1988. Grasslands. In *North American Terrestrial Vegetation,* ed. M. G. Barbour and W. D. Billings, 256–86. Cambridge: Cambridge University Press.

Sparks, K. L. 1972. Grazing behavior of bison and cattle on a shortgrass prairie. U.S. International Biological Program, Grassland Biome Technical Report No. 149. Fort Collins: Natural Resource Ecology Laboratory, Colorado State University.

Steelquist, R. U. 1998. *Field Guide to the North American Bison.* Seattle: Sasquatch.

Steinauer, G. 1999. Buffalo, the native grazer. NEBRASKAland 77(6): 10–19.

Strawn, S. A., and L. L. Wallace. 1994. Prairie dogs and ungulates: Competitors or mutual beneficiaries? Abstract. *Bulletin of the Ecological Society of America* 75(2 Supp.): 222.

Truett, J. C., M. Phillips, K. Kunkel, and R. Miller. 2001. Managing bison to restore biodiversity. *Great Plains Research* 11:123–44.

Turbak, G. 1995. *Pronghorn: Portrait of the American Antelope.* Flagstaff AZ: Northland.

Vanderhye, A. V. 1985. Interspecific nutritional facilitation: Do bison benefit from feeding on prairie dog towns? Master's thesis, Colorado State University.

Wilson, G. A., and C. Strobeck. 1998. Microsatellite analysis of genetic variation in woods and plains bison. In *Bison Ecology and Management in North America, Symposium Proceedings,* ed. L. I. Irby and J. E. Knight, 180–91. Bozeman: Montana State University.

Wuerther, G. 1998. Are cows just domestic bison? Behavioral and habitat differences between cattle and bison. In *Bison Ecology and Management in North America, Symposium Proceedings,* ed. L. I. Irby and J. E. Knight, 374–83. Bozeman: Montana State University.

Wydeven, A. P., and R. B. Dahlgren. 1985. Ungulate habitat relationships in Wind Cave National Park. *Journal of Wildlife Management* 49:805–13.

3. PRAIRIE DOGS AND THE AMERICAN WEST

For bibliographies of prairie dogs, see Hassien (1973, 1976), Clark (1986), and Knowles and Knowles (1994). More recent references appear in Hoogland (1994) and Graves (2001).

Audubon, J. J., and J. Bachman. 1854. *The Quadrupeds of North America*. 3 vols. Reprinted as *Audubon's Quadrupeds of North America*, 1 vol., Secaucus NJ: Wellfleet, 1989.

Bailey, V. 1905. *Biological Survey of Texas*. Biological Survey, North American Fauna No. 25. Washington DC: U.S. Department of Agriculture.

———1931. *Mammals of New Mexico*. Biological Survey, North American Fauna No. 53. Washington DC: U.S. Department of Agriculture.

Bartlett, J. R. 1854. *Personal Narrative of Explorations and Incidents in Texas, New Mexico, California, Sonora, and Chihuahua Connected with the United States and Mexican Boundary Commission*. New York: Appleton.

Biggins, D. E., B. J. Miller, L. Hanebury, R. Oakleaf, A. Farmer, R. Crete, and A. Dood. 1993. A technique for evaluating black-footed ferret habitat. In *Management of Prairie Dog Complexes for the Introduction of the Black-footed Ferret*, Biological Report 93, ed. J. L. Oldemeyer, D. E. Biggins, B. J. Miller, and R. Crete, 73–88. Washington DC: U.S. Fish & Wildlife Service.

Biodiversity Legal Foundation. 1999. *Recommended Conservation Measures for Protection and Recovery of the Black-tailed Prairie Dog* (Cynomys ludovicianus) *and Its Shortgrass Prairie Ecosystem*. Louisville CO: Biodiversity Legal Foundation.

———, and J. C. Sharps. 1994. Petition to classify the black-tailed prairie dog (*Cynomys ludovicianus*) as a Category 2 Candidate species pursuant to the Administrative Procedures Act and the intent of the Endangered Species Act (letter of October 21, 1994). Biodiversity Legal Foundation, Louisville CO.

———, the Predator Project, and J. C. Sharps. 1998. Petition to list the black-tailed prairie dog. Biodiversity Legal Foundation, Louisville CO.

Bishop, N., and J. Culbertson. 1976. Decline of prairie dog towns in southwestern North Dakota. *Journal of Range Management* 29:217–20.

Bonham, C. D., and J. S. Hannan. 1978. Blue grama and buffalograss patterns in and near a prairie dog town. *Journal of Range Management* 31:63–65.

———, and A. Lerwick. 1976. Vegetation changes induced by prairie dogs on shortgrass range. *Journal of Range Management* 29:221–25.

Carlson, D. C., and E. M. White. 1987. Effects of prairie dogs on mound soils. *Soil Science Society of America Journal* 51:389–93.

Casey, D. 1987. *The Friendly Prairie Dog*. New York: Dodd, Mead.

Ceballos, G., E. Mellink, and L. R. Hanebury. 1993. Distribution and conservation status of prairie dogs *Cynomys mexicanus* and *Cynomys ludovicianus* in Mexico. *Biological Conservation* 63:105–12.

Center for Native Ecosystems, Biodiversity Conservation Alliance, Southern Utah Conservation Alliance, American Lands Alliance, and Forest Guardians. 2002. White-tailed prairie dog: Petition to list the white-tailed prairie dog (*Cynomys leucurus*) as threatened or endangered under the Endangered Species Act, 16 USC 1531 et seq. as amended and for designation of critical habitat. Filed July 11, 2002, Paonia CO.

Cheatheam, L. K. 1977. Density and distribution of the black-tailed prairie dog in Texas. *Texas Journal of Science* 24:33–40.

Clark, T. W. 1968. Ecological roles of prairie dogs. *Wyoming Journal of Range Management* 261:102–4.

———. 1971. Towards a literature review of prairie dogs. *Wyoming Journal of Range Management* 286:29–44.

———. 1973. A field study of the ecology and ethology of the white-tailed prairie dog (*Cynomys leucurus*), with a model for *Cynomys* evolution. PhD diss., University of Wisconsin–Madison.

———. 1977. Ecology and ethology of the white-tailed prairie dog (*Cynomys leucurus*). *Milwaukee Public Museum Publications in Biology and Geology* 3.

———. 1979. The hard life of the prairie dog. *National Geographic* 156:270–81.

———. 1986. *Annotated Prairie Dog Bibliography 1973 to 1985*. BLM Wildlife Technical Bulletin No. 1. Billings MT: U.S. Bureau of Land Management; Helena MT: U.S. Fish & Wildlife Service.

———, T. M. Campbell III, D. G. Socha, and D. E. Casey. 1982. Prairie dog colony attributes and associated vertebrate species. *Great Basin Naturalist* 42:572–82.

———, D. K. Hinckley, and T. Rich, eds. 1989. *The Prairie Dog Ecosystem: Managing for Biological Diversity*. Technical Bulletin No. 2. Billings MT: U.S. Bureau of Land Management.

Collier, G. G. 1975. The Utah prairie dog: Abundance, distribution and habitat requirements. PhD diss., Utah State University.

———, and J. J. Spillett. 1975. Factors affecting the distribution of the Utah prairie dog, *Cynomys parvidens* (Sciuridae). *Southwestern Naturalist* 20:151–58.

Costello, D. F. 1970. *The World of the Prairie Dog*. Philadelphia: Lippincott.

Cully, J. F., Jr., and E. S. Williams. 2002. Interspecific comparisons of sylvatic plague in prairie dogs. *Journal of Mammalogy* 82:894–905.

Dale, H. F. 1947. Prairie dogs as pets. *Outdoor Nebraska* 24:22.

Davis, A. H. 1966. Winter activities of the black-tailed prairie dog in north-central Colorado. Master's thesis, Colorado State University.

Deisch, M. S., D. W. Uresk, and R. L. Linder. 1989. Effects of two prairie dog rodenticides on ground-dwelling invertebrates in western South Dakota. In *Proceedings Ninth Great Plains Wildlife Damage Control Workshop* (April 17–20, 1988, Fort Collins CO), ed. D. W. Uresk et al., 166–70. General Technical Report RM-171. Fort Collins CO: U.S. Forest Service.

Detling, J. K. 1998. Mammalian herbivores: Ecosystem-level effects in two grassland national parks. *Wildlife Society Bulletin* 26:438–48.

———, and A. D. Whicker. 1988. *Control of Ecosystem Processes by Prairie Dogs and Other Grassland Herbivores*. General Technical Report RM-154, pp. 23–29. Fort Collins CO: U.S. Forest Service.

Egoscue, H. J. 1975. The care, management and display of prairie dogs (*Cynomys* spp.). *International Zoo Yearbook* 154:45–48.

Fagerstone, K. A. 1979. Food habits of the black-tailed prairie dog, *Cynomys ludoviciana*. Master's thesis, University of Colorado.

———. 1982. A review of prairie dog diet and its variability among animals and colonies. In *Proceedings Fifth Great Plains Wildlife Damage Control Workshop* (October 13–15, 1981), pp. 178–84. Lincoln: Institute of Agriculture and Natural Resources, University of Nebraska.

FaunaWest. 1998. *Status of the Black and White-tailed Prairie Dog in Montana*. Helena: Montana Fish, Wildlife & Parks.

Fitzgerald, J. P., and R. P. Lechleitner. 1974. Observations on the biology of Gunnison's prairie dog in central Colorado. *American Midland Naturalist* 92:146–63.

Goodwin, H. T. 1995. The Pliocene–Pleistocene biogeographic history of prairie dogs genus *Cynomys* (Sciuridae). *Journal of Mammalogy* 76:100–122.

Graves, R. A. 2001. *The Prairie Dog: Sentinel of the Plains*. Lubbock: Texas Tech University Press.

Gunderson, H. L. 1978. Under and around a prairie dog town. *Natural History* 87(8): 56–67.

Halleron, A. F. 1972. The black-tailed prairie dog: Yesterday and today. *Great Plains Journal* 11:138–44.

Halpin, Z. T. 1983. Naturally occurring encounters between black-tailed prairie dogs (*Cynomys ludovicianus*) and snakes. *American Midland Naturalist* 109:50–54.

Hassien, F. D. 1973. Prairie dogs: A partial bibliography. In *Proceedings of the Black-footed Ferret and Prairie Dog Workshop*, ed. R. L. Lindner and C. N. Hillman, 178–205. Brookings: South Dakota State University.

———. 1976. Prairie dog bibliography. U.S. Department of the Interior, BLM *Technical Note* 279.

Haynie, M. L., R. A. Van Den Busche, J. L. Hoogland, and D. A. Gilbert. 2003. Parentage, multiple paternity, and breeding success in Gunnison's and Utah prairie dogs. *Journal of Mammalogy* 84:1244–53.

Hoogland, J. L. 1981. The evolution of coloniality in white-tailed and black-tailed prairie dogs (Sciuridae: *Cynomys leucurus* and *C. ludovicianus*). *Ecology* 62:252–72.

———. 1994. *The Black-tailed Prairie Dog: Social Life of a Burrowing Mammal*. Chicago: University of Chicago Press.

———. 2003. Sexual dimorphism in prairie dogs. *Journal of Mammalogy* 84:1254–66.

Johnson, K., T. Neville, and L. Pierce. 2003. *Remote Sensing Survey of Black-tailed Prairie Dog Towns in the Historical New Mexico Range*. New Mexico Natural Heritage Program Publication No. 03-GTR-248. Santa Fe: New Mexico Natural Heritage Program.

King, J. A. 1955. Social behavior, social organization, and population dynamics in a black-tailed prairie-dog town in the Black Hills of South Dakota. In *Contributions from the Laboratory of Vertebrate Biology* 67. Ann Arbor: University of Michigan.

Klatt, L. E., and D. Hein. 1978. Vegetative differences among active and abandoned towns of black-tailed prairie dogs (*Cynomys ludovicianus*). *Journal of Range Management* 31:315–17.

Knowles, C. J. 1982. Habitat affinity, populations, and population dynamics in a prairie dog town in the Black Hills of South Dakota. PhD diss., University of Montana.

———. 1987. Reproductive ecology of black-tailed prairie dogs in Montana. *Great Basin Naturalist* 47:202–6.

———. 1995. A summary of black-tailed prairie dog abundance and distribution on the central and northern Great Plains. Unpublished report to Defenders of Wildlife, Missoula MT.

———. 2002. *Status of White-tailed and Gunnison's Prairie Dogs*. Missoula MT: National Wildlife Federation; Washington DC: Environmental Defense.

———, J. D. Procter, and S. C. Forrest. 2002. Black-tailed prairie dog abundance and distribution in the Great Plains based on historic and contemporary information. *Great Plains Research* 12:219–54.

———, and P. R. Knowles. 1994. A review of black-tailed prairie dog literature in relation to rangelands administered by the Custer National Forest. Billings MT: U.S. Forest Service.

Korth, W. W. 1996. A new genus of prairie dog (Sciuridae, Rodentia) from the Miocene (Barstovian of Montana and Clarendonian of Nebraska) and the classification of Nearctic ground squirrels (Marmotini). *Transactions of the Nebraska Academy of Science* 23:109–13.

Lerwick, A. C. 1974. The effects of the black-tailed prairie dog on vegetative composition and their diet in relation to cattle. Master's thesis, Colorado State University.

Long, M. E. 1998. The vanishing prairie dog. *National Geographic* 193(4): 117–31.

Loughry, W. J. 1987. The dynamics of snake harassment by black-tailed prairie dogs. *Behaviour* 103:23–43.

Luce, R., ed. 2003. *A Multi-state Conservation Plan for the Black-tailed Prairie Dog*, Cynomys ludovicianus, *in the United States – an Addendum to the Black-tailed Prairie Dog Conservation Assessment and Strategy, November 3, 1999*. Washington DC: Wildlife Management Institute.

MacDonald, N. F., and S. E. Hygnstrom. 1991. Little dogs of the prairie. NEBRASKA*land* 69(5): 24–31.

Madson, J. 1968. Dark days in dogtown. *Audubon* 70:32–43.

Marcy, R. B. 1866. *Thirty Years of Army Life on the Border*. Reprint, Philadelphia: Lippincott, 1963.

Merriam, C. H. 1902. The prairie dog of the Great Plains. In *Yearbook of the United States Department of Agriculture, 1901*, 257–70. Washington DC: U.S. Department of Agriculture.

Miller, B. 1996. Delineating a new protected area in northern Chihuahua. *Wild Earth* 6(2): 14–16.

Miller, S., and J. Cully Jr. 2001. Conservation of black-tailed prairie dogs (*Cynomys ludovicianus*). *Journal of Mammalogy* 82:889–93.

Mulhern, D. W., and C. J. Knowles. 1997. Black-tailed prairie dog status and future conservation planning. In *Conserving Biodiversity on Native Rangelands* (Symposium Proceedings, Fort Robinson State Park NE, 1995), ed. D. W. Uresk et al., 19–29. General Technical Report RM-GTR-298. Fort Collins CO: U.S. Forest Service.

National Wildlife Federation. 1998. Petition to list the black-tailed prairie dog. Washington DC: National Wildlife Federation.

Nelson, E. W. 1919. Annual report to Chief of Biological Survey. In *U.S. Dept. Agriculture Annual; Rept. for 1919*, 275–98. Washington DC: U.S. Department of Agriculture.

O'Melia, M. E., F. Knopf, and J. Lewis. 1982. Some consequences of competition between prairie dogs and beef cattle. *Journal of Range Management* 35:580–85.

Osborn, B., and P. F. Allan. 1949. Vegetation of an abandoned prairie-dog town in tall grass prairie. *Ecology* 30:322–32.

Owings, D. H., and S. C. Owings. 1979. Snake-directed behavior by black-tailed prairie dogs. *Zietschrift für Tierpsychologie* 70:177–200.

Pfeiffer, E. W., L. N. Reinking, and J. D. Hamilton (1979). Some effects of food and water deprivation on metabolism in prairie dogs *Cynomys ludovicianus*. *Comparative Biochemistry and Physiology* 63:19–22.

Pizzimenti, J. J. 1975. Evolution of the prairie dog genus *Cynomys*. *Occasional Papers of the Museum of Natural History, University of Kansas* 39:1–73.

Potter, R. L., and R. M. Hansen. 1980. Early plant succession following removal of prairie dogs on shortgrass range. Abstract. *Proceedings of the Society of Range Management Annual Meeting* 33:41.

Predator Conservation Alliance. 2003. *Restoring the Prairie Dog Ecosystem of the Great Plains*. Bozeman MT: Predator Conservation Alliance.

Reading, R. P., S. R. Beissinger, J. J. Grensten, and T. W. Clark. 1989. Attributes of black-tailed prairie dog colonies in north-central Montana, with management recommendations for the conservation of biodiversity. In *The Prairie Dog Ecosystem: Managing for Biological Diversity*, ed. T. W. Clark, D. Hinkley, and T. Rich, 13–28. Montana BLM Wildlife Technical Bulletin No. 2. Billings MT: U.S. Bureau of Land Management.

Rosmarino, N. J. 2003. Comments re annual black-tailed prairie dog (*Cynomys ludovicianus*) status review information request. Letter of February 3, 2003, to U.S. Fish & Wildlife Service, Pierre SD.

———, E. Robertson, and D. Crawford. 2003. Comments re draft Grassland Species Conservation Plan. Letter of October 13, 2003, to Colorado Division of Wildlife, for Forest Guardians, Center for Native Ecosystems, and Rocky Mountain Animal Defense.

Seabloom, R. W., and P. W. Theisen. 1990. Breeding biology of the black-tailed prairie dog in North Dakota. *Prairie Naturalist* 22:65–74.

Scott, J. D. 1977. *Little Dogs of the Prairie*. New York: Putnam's.

Scott-Morales, L., E. Estrada, F. Chavez-Ramirez and M. Cotera. 2004. Continued decline in geographic distribution of the Mexican prairie dog. *Journal of Mammalogy*, 85:1095–1101.

Smith, N. 1979. Life in a prairie dog town. *National Wildlife* 17:38–39.

Smith, R. E. 1958. *Natural History of the Prairie Dog in Kansas*. Miscellaneous Publication No. 49. Lawrence: University of Kansas Natural History Museum.

Smith, W. J., S. L. Smith, J. G. deVilla, and E. C. Oppenheimer. 1976. The jump-yip display of the black-tailed prairie dog *Cynomys ludovicianus*. *Animal Behaviour* 24:609–21.

———, S. L. Smith, E. C. Oppenheimer, and J. G. deVilla. 1977. Vocalizations of the black-tailed prairie dog, *Cynomys ludovicianus*. *Animal Behaviour* 25:152–64.

———, and F. A. Ulmer. 1973. Behavior of a captive population of black-tailed prairie dogs: Annual cycles of social behavior. *Behaviour* 46:189–220.

Stephl, O. E. 1932. Battle between a prairie dog and a rattlesnake. *Journal of Mammalogy* 13:74–75.

Stockrahm, D. M. B., T. E. Olson, and E. K. Harper. 1993. Plant species in black-tailed prairie dog towns in Billings County, North Dakota. *Prairie Naturalist* 25:173–83.

Taylor, W. P., and J. V. G. Loftfield. 1924. Damage to range grasses by the Zuni prairie dog. *Bulletin of the U.S. Department of Agriculture* 1227:1–16.

Tileston, J. V., and R. R. Lechleitner. 1966. Some comparisons of the black-tailed and white-tailed prairie dogs in north-central Colorado. *American Midland Naturalist* 75:292–315.

Treviño-Villarreal, J., and W. E. Grant. 1998. Geographic range of the endangered Mexican prairie dog (*Cynomys mexicanus*). *Journal of Mammalogy* 79:1273–87.

Uresk, D. W. 1984. Black-tailed prairie dog food habits and forage relationships in western South Dakota. *Journal of Range Management* 38:466–68.

———, and A. Bjugstad. 1980. Prairie dogs as ecosystem regulators on the northern high plains. In *Proceedings of the Seventh North American Prairie Conference* (August 4–6, 1980), ed. C. Kucera. Springfield: Southwest Missouri State University.

———, G. L. Schenbeck, and J. T. Rourke, eds. 1997. *Conserving Biodiversity on Native Rangelands: Symposium Proceedings*. General Technical Note RM 298 (August 17). Fort Collins CO: U.S. Forest Service.

U.S. Fish & Wildlife Service. 1991. *Utah Prairie Dog Recovery Action Plan*. Denver: U.S. Fish & Wildlife Service.

———. 2000. 12-month finding for a petition to list the black-tailed prairie dog as threatened. *Federal Register* 65:5476–88.

Van Putten, M., and S. D. Miller. 1999. Prairie dogs: The case for listing. *Wildlife Society Bulletin* 27:113–20.

Vanderhoof, J. L., R. J. Robel, and K. E. Kemp. 1994. Numbers and extent of black-tailed prairie dog towns in Kansas. *Transactions of the Kansas Academy of Science* 97:36–43.

Virchow, D. R., and S. E. Hyngstrom. 2002a. Distribution and abundance of prairie dogs on the Great Plains: A historical perspective. *Great Plains Research* 12:197–218.

———. 2002b. Estimation of presettlement populations of the black-tailed prairie dog: A reply. *Great Plains Research* 12:255–60.

Vogel, S., C. P. Ellington Jr., and D. L. Kilgore Jr. 1973. Wind-induced ventilation of the burrow of the prairie dog, *Cynomys ludovicianus*. *Journal of Comparative Physiology* 85:1–14.

Waring, G. H. 1970. Sound communication of black-tailed, white-tailed and Gunnison's prairie dogs. *American Midland Naturalist* 83:167–84.

Weltzin, J. F., S. Archer, and R. K. Heitschmidt. 1997. Small-animal regulation of vegetation structure in a temperate savanna. *Ecology* 78:751–63.

———, S. L. Dowhawser, and R. K. Heitschmidt. 1997. Prairie dog effects on plant community structure in southern mixed-grass prairie. *Southwestern Naturalist* 42:251–58.

Whicker, A. D., and J. K. Detling. 1988. Ecological consequences of prairie dog disturbances. *BioScience* 38:778–85.

Wright-Smith, M. A. 1978. The ecology and social organization of *Cynomys parvidens* (Utah prairie dog) in southcentral Utah. Master's thesis, Indiana University.

Wuerther, G. 1995. Last chance for the prairie dog. *Wild Earth* 5(1): 21–25.

4. FERRETS, BADGERS, BOBCATS, AND COYOTES

Literature surveys of the coyote may be found in Beckoff (1974), Young and Jackson (1951), and Ryden (1979). References on ferrets may be found in Wood (1986) and in Miller, Reading, and Forrest (1996). American badger literature was provided by Long and Killingley (1983). Bobcat literature was summarized by Young (1958) and by Wassmer, Gruenther, and Layne (1988).

Anderson, E., S. C. Forrest, T. W. Clark, and L. Richardson. 1986. Paleobiology, biogeography, and systematics of the black-footed ferret, *Mustela nigripes* (Audubon and Bachman) 1851. *Great Basin Naturalist Memoirs* 8:11–62.

Anderson, S., and D. Inkley, eds. 1985. *Black-footed Ferret Workshop Proceedings*. Cheyenne: Wyoming Game & Fish Publications.

Beckoff, M. 1974. *A General Bibliography of the Coyote* (Canis latrans). Boulder CO: Coymar.

———. 1978. *Coyotes: Biology, Behavior and Management*. New York: Academic Press.

Cadieux, C. 1983. *Coyotes: Predators and Survivors*. New York: Stone Wall.

Cahalane, V. R. 1950. Badger-coyote "partnerships." *Journal of Mammalogy* 31:354–55.

Casey, D. E., J. DuWaldt, and T. W. Clark. 1986. Annotated bibliography of the black-footed ferret. *Great Basin Naturalist Memoirs* 8:185–208.

Clark, T. W. 1989. *Conservation Biology of the Black-footed Ferret*. Special Scientific Report No. 3. Philadelphia: Wildlife Preservation Trust International.

————. 1997. *Averting Extinction: Reconstructing Endangered Species Recovery*. New Haven CT: Yale University Press.

Dobie, J. F. 1949. *The Voice of the Coyote*. New York: Little, Brown.

Gipson, P. S. 1974. Food habits of coyotes in Arkansas. *Journal of Wildlife Management* 38:848–53.

Goldman, E. A. 1930. The coyote as archpredator. *Journal of Mammalogy* 11:325–35.

Goodrich, J. M., and S. W. Buskirk. 1998. Spacing and ecology of North American badgers (*Taxidea taxus*) in a prairie-dog (*Cynomys leucurus*) complex. *Journal of Mammalogy* 79:171–79.

Grady, W. 1994. *The Nature of Coyotes*. Vancouver: Douglas & McIntyre.

Henderson, F. R., P. Springer, and R. Adrian. 1969. *The Black-footed Ferret in South Dakota*. Technical Bulletin No. 4. Pierre: South Dakota Department of Game, Fish & Parks.

Hillman, C. N., R. L. Linder, and R. B. Dahlgren. 1979. Prairie dog distribution in areas inhabited by black-footed ferrets. *American Midland Naturalist* 102:185–87.

Knopf, F. L., and D. F. Balph. 1969. Badgers plug burrows to confine prey. *Journal of Mammalogy* 50:635–36.

Leydet, F. 1988. *The Coyote: Defiant Songdog of the West*. Norman: University of Oklahoma Press.

Linder, R. L., and C. N. Hillman, eds. 1973. *Proceedings of the Black-footed Ferret and Prairie Dog Workshop*. Brookings: South Dakota State University.

Line, L. 1997. Phantom of the plains: The continuing saga of the black-footed ferret. *Wildlife Conservation* 100:20–27.

Long, C. A., and C. A. Killingley. 1983. *The Badgers of the World*. Springfield IL: C. C. Thomas.

Lowie, R. H. 1918. Myths and traditions of the Crow Indians. *Anthropological Papers of the American Museum of Natural History* 25(1): 1–308.

Miller, B. 1996. Delineating a new protected area in northern Chihuahua. *Wild Earth* 6(2): 14–16.

————, R. P. Reading, and S. Forrest. 1996. *Prairie Night: Black-footed Ferrets and the Recovery of Endangered Species*. Washington DC: Smithsonian Institution Press.

Murie, A. 1940. *Ecology of the Coyote of the Yellowstone*. Fauna Series No. 4. Washington DC: U.S. Government Printing Office.

National Wildlife Federation. 1998. Petition to list the black-tailed prairie dog. Washington DC: National Wildlife Federation.

Neal, E. G. 1996. *The Natural History of Badgers*. New York: Facts on File.

Oldemeyer, J. L., D. E. Biggins, B. J. Miller, and R. Crete, eds. 1993. *Management of Prairie Dog Complexes for the Introduction of the Black-footed Ferret*. Biological Report No. 93. Washington DC: U.S. Fish & Wildlife Service.

Powell, J. W. 1881. Sketch of the mythology of the North American Indians. *Annual Report of the Bureau of American Ethnology, 1878–1880* 1:19–56.

Powell, R. A. 1982. Prairie dog coloniality and black-footed ferrets. *Ecology* 63:1967–68.

Pringle, L. 1977. *The Controversial Coyote: Predation, Politics and Ecology.* New York: Harcourt Brace Jovanovich.

Robinson, W. B. 1961. Population changes of carnivores in some coyote-control areas. *Journal of Mammalogy* 42:510–15.

Roosevelt, T. 1900. *The Wilderness Hunter.* 2 vols. New York: Putnam.

Rosmarino, N. J. 2003. Comments re annual black-tailed prairie dog (*Cynomys ludovicianus*) status review information request. Letter of February 3, 2003, to U.S. Fish & Wildlife Service, Pierre SD.

Ryden, H. 1979. *God's Dog: A Celebration of the North American Coyote.* New York: Coward, McCann & Geohagen.

Seton, E. T. 1929. *Lives of Game Animals.* 4 vols. Boston: Chas. Branford.

Silver, J. 1928. Badger activities in prairie-dog control. *Journal of Mammalogy* 9:63.

Sperry, C. C. 1941. Food habits of the coyote. (U.S. Department of the Interior) *Wildlife Research Bulletin* 4:9–11.

Stewart, D. 1999. Caught in a dog fight. *National Wildlife* 37(4): 35–39.

U.S. Fish & Wildlife Service. 1988. *Black-footed Ferret Recovery Plan.* Denver: U.S. Fish & Wildlife Service.

Wassmer, D. A., D. D. Gruenther, and J. N. Layne. 1988. Ecology of the bobcat in south-central Florida. *Bulletin of the Florida State Museum, Biological Sciences* 33:159–228.

Wood, S. L., ed. 1986. The black-footed ferret. *Great Basin Naturalist Memoirs* 8:1–208.

Young, S. P. 1958. *The Bobcat of North America.* Washington DC: Wildlife Management Institute.

Young, S. P., and H. H. T. Jackson. 1951. *The Clever Coyote.* Lincoln: University of Nebraska Press.

5. FREE-LOADERS AND HANGERS-ON

Literature reviews of the burrowing owl can be found in Haug, Millsap, and Martel (1993), Dechant et al. (1999), and Johnsgard (2002). For other associated birds see the references for Chapters 6 and 7. For recent literature on the swift fox and kit fox, see the online bibliography at http://www.wildlifer.com.

Agnew, W. D. 1983. Flora and fauna associated with prairie dog ecosystems. Master's thesis, Colorado State University.

———, D. W. Uresk, and R. M. Hansen. 1986. Flora and fauna associated with prairie dog colonies and adjacent ungrazed mixed-grass prairie in western South Dakota. *Journal of Range Management* 39:135–39.

Anderson, D. E., T. R. Laurion, J. R. Cary, R. S. Sikes, and E. M. Gese. 1998. Ecology of swift fox in southeastern Colorado. Abstract. In *Swift Fox Symposium: Ecology and Conservation of Swift Foxes in a Changing World.*

Arrowwood, P. C., C. A. Finley, and B. C. Thompson. 2001. Analyses of burrowing owl populations in New Mexico. *Journal of Raptor Research* 35:362–70.

Avery, S. R. 1990. Vocalizations and behavior of the swift fox (*Vulpes velox*). Master's thesis, University of Northern Colorado.

Bishop, N. G., and J. L. Culbertson. 1976. Decline of prairie dog colonies in southwestern North Dakota. *Journal of Range Management* 29:217–20.

Boddiger, M. J., and A. Lerwick. 1976. Vegetation changes induced by prairie dogs on shortgrass range. *Journal of Range Management* 29:221–25.

Bright, J., A. McIntire, and J. Sneva. 1998. Prairie dogs and short-grass prairie ecosystems. *Arizona Wildlife Views* 41(5): 10–12.

Butts, K. O. 1973. Life history and habitat requirements of burrowing owls in western Oklahoma. Master's thesis, Oklahoma State University.

———, and J. C. Lewis. 1982. The importance of prairie dog towns to burrowing owls in Oklahoma. *Proceedings of the Oklahoma Academy of Science* 62:46–52.

Cameron, M. W. 1984. The swift fox (*Vulpes velox*) on the Pawnee National Grassland: Its food habits, population dynamics, and ecology. Master's thesis, University of Northern Colorado.

Campbell, T. M., III, and T. W. Clark. 1981. Colony characteristics and vertebrate associates of white-tailed and black-tailed prairie dogs in Wyoming. *American Midland Naturalist* 105:269–76.

Ceballos, G., J. Pacheco, and R. List. 1999. Influence of prairie dogs (*Cynomys ludovicianus*) on habitat heterogeneity and mammalian diversity in Mexico. *Journal of Arid Environments* 41:161–72.

Clayton, K. M., and J. K. Schmutz. 1999. Is the decline of burrowing owls *Speotyto cunicularia* in prairie Canada linked to changes in Great Plains ecosystems? *Bird Conservation International* 9:163–85.

Conrad, S. T., J. L. Dose, and D. N. Svingen. No date. Burrowing owl association with prairie dog towns in the southern Great Plains. Unpublished manuscript, U.S. Forest Service, Springfield CO.

Covell, D. F. 1992. Ecology of the swift fox (*Vulpes velox*) in southeastern Colorado. Master's thesis, University of Wisconsin–Madison.

Cutter, W. L. 1958. Food habits of the swift fox in northern Texas. *Journal of Mammalogy* 39:527–32.

Dechant, D. J., M. L. Sondreal, D. H. Johnson, L. D. Igl, C. M. Goldade, M. P. Nenneman, and B. R. Euliss. 1999 (rev. 2001). *Effects of Management Practices on Grassland Birds: Burrowing Owl.* Jamestown ND: Northern Prairie Wildlife Research Center.

Desmond, M. J., and J. A. Savidge. 1995. Spatial patterns of burrowing owl (*Speotyto cunic-*

ularia) nests within black-tailed prairie dog (*Cynomys ludovicianus*) towns. *Canadian Journal of Zoology* 73:1375–79.

———. 1996. Factors influencing burrowing owl (*Speotyto cunicularia*) nest densities and numbers in western Nebraska. *American Midland Naturalist* 136:143–48.

———, and K. M. Eskridge. 2000. Correlations between burrowing owl and black-tailed prairie dog declines: A 7-year analysis. *Journal of Wildlife Management* 64:1067–75.

Dragoo, J. W., J. R. Choate, T. L. Yates, and T. P. O'Farrell. 1990. Evolutionary and taxonomic relationships among North American arid-land foxes. *Journal of Mammalogy* 71:318–32.

Green, G. A., and R. G. Anthony. 1989. Nesting success and habitat relationships of burrowing owls in the Columbia Basin, Oregon. *Condor* 91:347–54.

Grossman, J. 1987. A prairie dog companion. *Audubon* 89:52–67.

Halpin, Z. T. 1983. Naturally occurring encounters between black-tailed prairie dogs (*Cynomys ludovicianus*) and snakes. *American Midland Naturalist* 109:50–54.

Haug, E. A, B. A. Millsap, and M. S. Martell. 1993. Burrowing owl (*Athene cunicularia*). In *The Birds of North America*, no. 61, ed. A. Poole and F. Gill. Philadelphia: Academy of Natural Sciences; Washington DC: American Ornithologists' Union.

Herrero, S., C. Schroeder, and M. Scott-Brown. 1986. Are Canadian foxes swift enough? *Biological Conservation* 36:159–67.

Holroyd, G., R. Rodriquez-Estrella, and S. R. Sheffield. 2001. Conservation of the burrowing owl in western North America: Issues, challenges and recommendations. *Journal of Raptor Research* 35:399–407.

Hughes, A. J. 1993. Breeding density and habitat preference of the burrowing owl in northeastern Colorado. Master's thesis, Colorado State University.

James, P. C., and R. H. M. Espie. 1997. Current status of the burrowing owl in North America: An agency survey. In *The Burrowing Owl, Its Biology and Management, Including the Proceedings of the First International Burrowing Owl Symposium*, Journal of Raptor Research Report 9, ed. J. L. Lincer and K. Steenhof, 3–5. Boise ID: Raptor Research Foundation.

Johnsgard, P. A. 2002. *North American Owls: Biology and Natural History*. Washington DC: Smithsonian Institution Press.

Harrison, R. L. 2003. Swift fox demography, movements, denning and diet in New Mexico. *Southwestern Naturalist* 48:261–73.

Hines, T. D. 1980. An ecological study of *Vulpes velox* in Nebraska. Master's thesis, University of Nebraska.

———, and R. M. Case. 1991. Diet, home range, movements, and activity periods of swift fox in Nebraska. *Prairie Naturalist* 23:131–38.

Jackson, V. L., and J. R. Choate. 2000. Dens and den sites for the swift fox, *Vulpes velox*. *Southwestern Naturalist* 45:212–20.

Jones, J. K., C. Jones, R. R. Hollander, and R. W. Manning. 1987. *The Swift Fox in Texas.* Austin: Texas Parks & Wildlife.

Kamler, J. F., W. B. Ballard, E. B. Fish, P. R. Lemons, K. Monte, and C. C. Perchellet. 2003. Habitat use, home ranges and survival of swift foxes in a fragmented landscape: Conservation implications. *Journal of Mammalogy* 84:989–95.

Kilgore, D. L., Jr. 1969. An ecological study of the swift fox (*Vulpes velox*) in the Oklahoma panhandle. *American Midland Naturalist* 81:512–34.

Kitchen, A. M., E. M. Gese, and E. R. Schauster. 1999. Resource partitioning between coyotes and swift foxes: Space, time, and diet. *Canadian Journal of Zoology* 77:1645–56.

Klatt, I. E., and D. Hein. 1978. Vegetative differences among active and abandoned towns of black-tailed prairie dogs (*Cynomys ludovicianus*). *Journal of Range Management* 31:315–17.

Klute, D. S., A. W. Ayers, M. T. Green, W. H. Howe, S. L. Jones, J. A. Shaffer, S. R. Sheffield, and T. S. Zimmerman. 2003. *Status Assessment and Conservation Plan for the Western Burrowing Owl in the United States.* Biological Technical Publication FWS/BTP R6001–2003. Washington DC: U.S. Fish & Wildlife Service.

Konrad, P. M., and D. S. Gilmer. 1984. Observations on the nesting ecology of burrowing owls in central North Dakota. *Prairie Naturalist* 16:129–30.

Korphany, N. M., L. W. Ayers, S. H. Anderson, and D. B. McDonald. 2001. A preliminary assessment of burrowing owl population status in Wyoming. *Journal of Raptor Research* 35:337–43.

Kotliar, N. B. 2000. Application of the new keystone-species concept to prairie dogs: How well does it work? *Conservation Biology* 14:1715–21.

———, B. W. Baker, A. D. Whicker, and G. Plumb. 1999. A critical review of assumptions about the prairie dog as a keystone species. *Environmental Management* 24:177–92.

Lincer, J. L., and K. Steenhof, eds. 1997. *The Burrowing Owl, Its Biology and Management, Including the Proceedings of the First International Burrowing Owl Symposium.* Journal of Raptor Research Report 9. Boise ID: Raptor Research Foundation.

Loughry, W. J. 1987. The dynamics of snake harassment by black-tailed prairie dogs. *Behaviour* 103:23–43.

MacCracken, J. G., D. W. Uresk, and R. M. Hansen. 1985. Vegetation and soils of burrowing owl nest sites in Conata basin, South Dakota. *Condor* 87:152–54.

Maldonado, J. E., M. Cotera, E. Geffen, and R. K. Wayne. 1997. Relationships of the endangered Mexican kit fox (*Vulpes macrotis zinseri*) to North American arid-land foxes based on mitochondrial DNA sequence data. *Southwestern Naturalist* 42:460–70.

Manzano-Fischer, P., R. List, and G. Ceballos. 1999. Grassland birds in prairie-dog towns in northwestern Chihuahua, Mexico. *Studies in Avian Biology* 19:263–71. In *Ecology and Conservation of Grassland Birds of the Western Hemisphere*, ed. P. D. Vickery and J. R. Herkert. Camarillo CA: Cooper Ornithological Society.

Maximilian, Prince of Wied. 1906. *Travels in the Interior of North America, 1832–1834.*

English translation, *Early Western Travels*, ed. R. G. Twaites, vols. 22–24 (Cleveland: Arthur H. Clark, 1904–1907).

Miller, B., G. Ceballos, and R. P. Reading. 1994. The prairie dog and biotic diversity. *Conservation Biology* 8:677–81.

———, R. Reading, J. Hoogland, T. Clark, G. Ceballos, R. List, S. Forrest, L. Hanebury, P. Manzano, J. Pacheco, and D. Uresk. 2000. The role of prairie dogs as a keystone species: Response to Stapp. *Conservation Biology* 14:318–21.

Murphy, R. K., K. W. Hasselblad, C. D. Grondahl, J. G. Sidle, R. E. Martin, and D. W. Freed. 2001. Status of the burrowing owl in North Dakota. *Journal of Raptor Research* 35:322–30.

Olson, T. L., and F. G. Lindzey. 2002. Swift fox survival and production in southeastern Wyoming. *Journal of Mammalogy* 83:199–206.

Orth, P. B., and P. L. Kennedy. 2001. Do land-use patterns influence nest-site selection by burrowing owls (*Athene cunicularia hypugaea*) in northwestern Colorado? *Canadian Journal of Zoology* 79:1038–45.

Pechacek, P., F. G. Lindzey, and S. H. Anderson. 2000. Home range size and spatial organization of swift fox *Vulpes velox* (Say, 1823) in southeastern Wyoming. *Zeitschrift für Saugetierkunde* 65:209–15.

Peek, M., ed. 2002. *Swift Fox Conservation Team Annual Report*. Emporia: Kansas Department of Wildlife & Parks.

Petraitis, P. S., R. E. Latham, and R. A. Niesenbaum. 1989. The maintenance of species diversity by disturbance. *Quarterly Review of Biology* 64:292–419.

Pezzolesi, L. S. 1994. The western burrowing owl: Increasing prairie dog abundance, foraging theory, and nest site fidelity. Master's thesis, Texas Tech University.

Poulin, R. G. 2003. Relationships between burrowing owls (*Athene cunicularia*), small mammals, and agriculture. PhD diss., University of Saskatchewan.

Restani, M., L. R. Rau, and D. L. Flath. 2001. Nesting ecology of burrowing owls occupying black-tailed prairie dog towns in southeastern Montana. *Journal of Raptor Research* 35:296–303.

Reading, R. P., S. R. Beissinger, J. J. Grensten, and T. W. Clark. 1989. Attributes of black-tailed prairie dog colonies in north-central Montana, with management recommendations for the conservation of biodiversity. In *The Prairie Dog Ecosystem: Managing for Biological Diversity*, ed. T. W. Clark, D. Hinkley, and T. Rich, 13–28. Montana BLM Wildlife Technical Bulletin No. 2. Billings MT: U.S. Bureau of Land Management.

Rohwer, S. A., and D. L. Kilgore Jr. 1973. Interbreeding in the arid-land foxes, *Vulpes velox* and *Vulpes macrotis*. *Systematic Zoology* 22:157–65.

Schauster, E. R., E. M. Gese, and A. M. Kitchen. 2002. Population ecology of swift foxes (*Vulpes velox*) in southeastern Colorado. *Canadian Journal of Zoology* 80:307–19.

Schmitt, C. G. 2000. *Swift Fox Conservation Report*. 1999 annual report. Albuquerque: New Mexico Department of Game & Fish.

Schmutz, J. K., G. Wood, and D. Wood. 1991. Spring and summer prey of burrowing owls in Alberta. *Blue Jay* 49:93–97.

Severe, D. S. 1977. Revegetation of black-tailed prairie dog mounds on shortgrass prairie in Colorado. Master's thesis, Colorado State University.

Shackford, J. S., and J. D. Tyler. 1991. *Vertebrates Associated with Black-tailed Prairie Dog Colonies in Oklahoma*. Oklahoma City: Non-game Program, Oklahoma Department of Wildlife Conservation.

Sharps, J. C., and D. W. Uresk. 1990. Ecological review of black-tailed prairie dogs and associated species in western South Dakota. *Great Basin Naturalist* 50:339–45.

Sheffield, S. R. 1997. Current status, distribution and conservation of the burrowing owl (*Speotyto cunicularia*) in midwestern and western North America. In *Biology and Conservation of Owls of the Northern Hemisphere*, ed. J. R. Duncan, D. H. Johnson, and T. H. Nicholls, 399–407. General Technical Report NC-190. St. Paul MN: U.S. Forest Service.

———, and M. Howery. 2001. Current status, distribution and conservation of the burrowing owl in Oklahoma. *Journal of Raptor Research* 35:352–56.

Shyry, D. T., T. I Wellicome, J. K. Schmutz, G. L. Erickson, D. L. Scobie, R. E. Russell, and R. G. Martin. 2001. Burrowing owl population trend surveys in southern Alberta, 1991–2000. *Journal of Raptor Research* 35:310–15.

Sidle, J. G., M. Ball, T. Byer, J. J. Chynoweth, G. Foli, R. Hodorff, G. Moravek, R. Peterson, and D. N. Svingen. 2001. Occurrence of burrowing owls in black-tailed prairie dog colonies on Great Plains national grasslands. *Journal of Raptor Research* 35:316–21.

Sovada, M. A., C. C. Roy, J. B. Bright, and J. R. Gillis. 1998. Causes and rates of mortality of swift foxes in western Kansas. *Journal of Wildlife Management* 62:1300–1306.

———, C. C. Roy, and D. J. Telesco. 2001. Seasonal food habitats of swift fox (*Vulpes velox*) in cropland and rangeland landscapes in western Kansas. *American Midland Naturalist* 145:101–11.

———, and L. Carbyn, eds. 2003. *Ecology and Management of Swift Foxes in a Changing World*. Regina SK: Canadian Plains Research Center, University of Regina.

Stapp, P. 1998. A reevaluation of the role of prairie dogs in Great Plains grasslands. *Conservation Biology* 12:1253–59.

Stromberg, M. R., and M. S. Boyce. 1986. Systematics and conservation of the swift fox, *Vulpes velox*, in North America. *Biological Conservation* 35:97–110.

Thompson, C. D., and S. H. Anderson. 1988. Foraging behavior and food habits of burrowing owls in Wyoming. *Prairie Naturalist* 20:23–28.

Tyler, J. D. 1968. Distribution and vertebrate associates of the black-tailed prairie dog in Oklahoma. PhD diss., University of Oklahoma.

———. Additional vertebrates in a Jackson County prairie dog town. Unpublished manuscript.

————. 1970. Vertebrates in a prairie dog town. *Proceedings of the Oklahoma Academy of Science* 50:110–13.

Uresk, D. W., and J. C. Sharps. 1986. Denning habitat and diet of the swift fox in western South Dakota. *Great Basin Naturalist* 46:249–53.

U.S. Geological Survey. 1998. *Swift Fox Symposium: Ecology and Conservation of Swift Foxes in a Changing World.* Abstracts. Available at the Northern Prairie Wildlife Research Center home page: http://www.npwrc.usgs.gov/resource/1998/swiftfox/swift fox.htm

VerCauteren, T. L., S. W. Gillihan, and S. W. Hutchings. 2001. Distribution of burrowing owls on public and private lands in Colorado. *Journal of Raptor Research* 35:357–61.

Walker, J. R. 1917. The Sun Dance and other ceremonies of the Oglala Division of the Dakota Teton. *Anthropological Papers of the American Museum of Natural History* 16(2).

Warnock, R. G., and P. C. James. 1997. Habitat fragmentation and burrowing owls (*Speotyto cunicularia*) in Saskatchewan. In *Biology and Conservation of Owls of the Northern Hemisphere,* ed. J. R. Duncan, D. H. Johnson, and T. H. Nicholls, 477–86. General Technical Report NC-190. St. Paul MN: U.S. Forest Service.

Wedgwood, J. A. 1976. Burrowing owls in south-central Saskatchewan. *Blue Jay* 34:26–44.

Wellicome, T. I., and G. L. Holroyd, eds. 2001. Proceedings of the 2nd International Burrowing Owl Symposium. *Journal of Raptor Research* 35:269–407.

Wilcomb, M. J. 1954. A study of prairie dog burrow systems and the ecology of their arthropod inhabitants in central Oklahoma. PhD diss., University of Oklahoma.

Zarn, M. 1974. *Burrowing owl* (Speotyto cunicularia hypugaea). Habitat Management Series for Endangered Species, Technical Note 250. Denver: U.S. Bureau of Land Management.

6. OTHER HIGH PLAINS WILDLIFE

For literature reviews of the pronghorn, see Turbak (1995), Wormer (1969), and Einarsen (1948). For general literature reviews of grassland birds, see Knopf (1996b) and Johnsgard (2001); for literature surveys of individual grassland bird species, see Dechant et al. (1999–2002).

Agnew, W. D. 1983. Flora and fauna associated with prairie dog ecosystems. Master's thesis, Colorado State University.

————, D. W. Uresk, and R. M. Hansen. 1986. Flora and fauna associated with prairie dog colonies and adjacent ungrazed mixed-grass prairie in western South Dakota. *Journal of Range Management* 39:135–39.

Allen, J. N. 1980. The ecology and behavior of the long-billed curlew in southeastern Washington. *Wildlife Monographs* 73:1–64.

Banko, V. A., J. H. Shaw, and D. M. Leslie Jr. 1999. Birds associated with black-tailed prairie dog colonies in southern shortgrass prairie. *Southwestern Naturalist* 44:484–89.

Beason, R. C. 1995. Horned lark (*Eremophila alpestris*). In *The Birds of North America*, no. 195, ed. A. Poole and F. Gill. Philadelphia: Birds of North America; Washington DC: American Ornithologists' Union.

Bicak, T. K. 1977. Some eco-ethological aspects of a breeding population of long-billed curlews (*Numenius americanus*) in Nebraska. Master's thesis, University of Nebraska–Omaha.

Boddiger, M. J., and A. Lerwick. 1976. Vegetation changes induced by prairie dogs on shortgrass range. *Journal of Range Management* 29:221–25.

Bright, J., A. McIntire, and J. Sneva. 1998. Prairie dogs and short-grass prairie ecosystems. *Arizona Wildlife Views* 41(5): 10–12.

Butterfield, J. D. 1969. Nest-site requirements of the lark bunting in Colorado. Master's thesis, Colorado State University.

Cable, T. T., S. Seltman, and K. J. Cook. 1996. *Birds of Cimarron National Grassland*. General Technical Report RM-GTR-281. Fort Collins CO: U.S. Forest Service.

Campbell, T. M., III, and T. W. Clark. 1981. Colony characteristics and vertebrate associates of white-tailed and black-tailed prairie dogs in Wyoming. *American Midland Naturalist* 105:269–76.

Ceballos, G., J. Pacheco, and R. List. 1999. Influence of prairie dogs (*Cynomys ludovicianus*) on habitat heterogeneity and mammalian diversity in Mexico. *Journal of Arid Environments* 41:161–72.

Cochrane, J. F. 1983. Long-billed curlew habitat and land-use relationships in western Wyoming. Master's thesis, University of Wyoming.

Colwell, M. A., and J. R. Jehl Jr. 1994. Wilson's phalarope (*Phalaropus tricolor*). In *The Birds of North America*, no. 83, ed. A. Poole and F. Gill. Philadelphia: Birds of North America; Washington DC: American Ornithologists' Union.

Connelly, J. W., M. W. Gratson, and K. P. Reese. 1998. Sharp-tailed grouse (*Tympanuchus phasianellus*). In *The Birds of North America*, no. 354, ed. A. Poole and F. Gill. Philadelphia: Birds of North America; Washington DC: American Ornithologists' Union.

Coppock, D. L., J. K. Detling, J. L. Dodd, and M. I. Dyer. 1980. Bison–prairie dog–plant interactions in Wind Cave National Park, South Dakota. *Proceedings of Conference on Science re the National Parks* 12:184.

Creighton, P. D. 1971. *Nesting of the Lark Bunting in North-central Colorado*. Grassland Biome Technical Report 29. Denver: U.S. International Biological Program.

Dechant, D. J., M. L. Sondreal, D. H. Johnson, L. D. Igl, C. M. Goldade, M. P. Nenneman, and B. R. Euliss. 1998a (rev. 2001). *Effects of Management Practices on Grassland Birds: Baird's Sparrow*. Jamestown ND: Northern Prairie Wildlife Research Center.

———. 1998b (rev. 2001). *Effects of Management Practices on Grassland Birds: Chestnut-collared Longspur*. Jamestown ND: Northern Prairie Wildlife Research Center.

———. 1998c (rev. 2001). *Effects of Management Practices on Grassland Birds: Clay-colored Sparrow*. Jamestown ND: Northern Prairie Wildlife Research Center.

———. 1998d (rev. 2001). *Effects of Management Practices on Grassland Birds: Grasshopper Sparrow.* Jamestown ND: Northern Prairie Wildlife Research Center.

———. 1998e (rev. 2001). *Effects of Management Practices on Grassland Birds: Loggerhead Shrike.* Jamestown ND: Northern Prairie Wildlife Research Center.

———. 1998f (rev. 2001). *Effects of Management Practices on Grassland Birds: Marbled Godwit.* Jamestown ND: Northern Prairie Wildlife Research Center.

———. 1998g (rev. 2001). *Effects of Management Practices on Grassland Birds: Mountain Plover.* Jamestown ND: Northern Prairie Wildlife Research Center.

———. 1998h (rev. 2001). *Effects of Management Practices on Grassland Birds: Short-eared Owl.* Jamestown ND: Northern Prairie Wildlife Research Center.

———. 1998i (rev. 2001). *Effects of Management Practices on Grassland Birds: Sprague's Pipit.* Jamestown ND: Northern Prairie Wildlife Research Center.

———. 1998j (rev. 2001). *Effects of Management Practices on Grassland Birds: Willet.* Jamestown ND: Northern Prairie Wildlife Research Center.

———. 1998k (rev. 2001). *Effects of Management Practices on Grassland Birds: Wilson's Phalarope.* Jamestown ND: Northern Prairie Wildlife Research Center.

———. 1999a (rev. 2001). *Effects of Management Practices on Grassland Birds: Lark Bunting.* Jamestown ND: Northern Prairie Wildlife Research Center.

———. 1999b (rev. 2001). *Effects of Management Practices on Grassland Birds: Lark Sparrow.* Jamestown ND: Northern Prairie Wildlife Research Center.

———. 1999c (rev. 2001). *Effects of Management Practices on Grassland Birds: Long-billed Curlew.* Jamestown ND: Northern Prairie Wildlife Research Center.

———. 1999d (rev. 2001). *Effects of Management Practices on Grassland Birds: McCown's Longspur.* Jamestown ND: Northern Prairie Wildlife Research Center.

———. 1999e (rev. 2001). *Effects of Management Practices on Grassland Birds: Western Meadowlark.* Jamestown ND: Northern Prairie Wildlife Research Center.

———. 2000a (rev. 2001). *Effects of Management Practices on Grassland Birds: Horned Lark.* Jamestown ND: Northern Prairie Wildlife Research Center.

———. 2000b (rev. 2001). *Effects of Management Practices on Grassland Birds: Upland Sandpiper.* Jamestown ND: Northern Prairie Wildlife Research Center.

———. 2000c (rev. 2001). *Effects of Management Practices on Grassland Birds: Vesper Sparrow.* Jamestown ND: Northern Prairie Wildlife Research Center.

Dinsmore, S. 2001. Population biology of mountain plover in southern Phillips County, Montana. PhD diss., Colorado State University.

Dolby, C. C. 1999. The national grasslands and disappearing biodiversity: Can the prairie dog save us from an ecological desert? *Environmental Law:* 29:213–34.

Dugger, B. D., and K. N. Dugger. 2002. Long-billed curlew (*Numenius americana*). In *The Birds of North America*, no. 628, ed. A. Poole and F. Gill. Philadelphia: Birds of North America; Washington DC: American Ornithologists' Union.

Einarsen, A. S. 1948. *The Pronghorn Antelope and Its Management*. Washington DC: Wildlife Management Institute.

Ellison, A. E., and C. M. White. 2001. Breeding biology of mountain plovers (*Charadrius montanus*) in the Uinta Basin. *Western North American Naturalist* 61:223–28.

Felske, B. E. 1971. The population dynamics and productivity of McCown's longspur at Matador, Saskatchewan. Master's thesis, University of Saskatchewan.

Fishbein, S. 1989. *Yellowstone Country: The Enduring Wonder*. Washington DC: National Geographic Society.

Fitzner, J. N. 1978. The ecology and behavior of the long-billed curlew (*Numenius americanus*) in southeastern Washington. PhD diss., Washington State University.

Forsythe, D. M. 1972. Observations on the nesting biology of the long-billed curlew. *Great Basin Naturalist* 32:88–90.

Giesen, K. M. 1998. Lesser prairie-chicken (*Tympanuchus pallidicinctus*). In *The Birds of North America*, no. 364, ed. A. Poole and F. Gill. Philadelphia: Birds of North America; Washington DC: American Ornithologists' Union.

Gratto-Trevor, C. 2000. Marbled godwit (*Limosa fedoa*). In *The Birds of North America*, no. 492, ed. A. Poole and F. Gill. Philadelphia: Birds of North America; Washington DC: American Ornithologists' Union.

Graul, W. D. 1973. Adaptive aspects of the mountain plover social system. *Living Bird* 12:69–94.

———. 1975. Breeding biology of the mountain plover. *Wilson Bulletin* 87:6–31.

Green, M. T., P. E. Lowther, S. L. Jones, S. K. Davis, and B. C. Dale. 2002. Baird's sparrow (*Ammodramus bairdii*). In *The Birds of North America*, no. 638, ed. A. Poole and F. Gill. Philadelphia: Birds of North America; Washington DC: American Ornithologists' Union.

Greer, R. D. 1988. Effects of habitat structure and productivity on grassland birds. PhD diss., University of Wyoming.

———, and S. H. Anderson. 1989. Relationships between population demography of McCown's longspurs and habitat resources. *Condor* 91:609–19.

Hill, D. P., and L. K. Gould. 1997. Chestnut-collared longspur (*Calcarius ornatus*). In *The Birds of North America*, no. 288, ed. A. Poole and F. Gill. Philadelphia: Birds of North America; Washington DC: American Ornithologists' Union.

Houston, C. S., and D. E. Bowen Jr. 2001. Upland sandpiper (*Bartramia longicauda*). In *The Birds of North America*, no. 580, ed. A. Poole and F. Gill. Philadelphia: Birds of North America; Washington DC: American Ornithologists' Union.

Jamison, B. E., J. A. Dechant, D. H. Johnson, L. D. Igl, C. M. Goldade, and B. R. Euliss. 2002. *Effects of Management Practices on Grassland Birds: Lesser Prairie-chicken*. Jamestown ND: Northern Prairie Wildlife Research Center.

Johnsgard, P. A. 1979. *Birds of the Great Plains: Breeding Species and Their Distribution*. Lincoln: University of Nebraska Press.

————. 2001. *Prairie Birds: Fragile Splendor in the Great Plains*. Lawrence: University Press of Kansas.

————. 2002. *Grassland Grouse and Their Conservation*. Washington DC: Smithsonian Institution Press.

Kingery, H. E., ed. 1998. *Colorado Breeding Bird Atlas*. Denver: Colorado Bird Partnership and Colorado Division of Wildlife.

Klatt, I. E., and D. Hein. 1978. Vegetative differences among active and abandoned towns of black-tailed prairie dogs (*Cynomys ludovicianus*). *Journal of Range Management* 31:315–17.

Knick, S. T., and J. T. Rotenberry. 1995. Landscape characteristics of fragmented shrub-steppe habitats and breeding passerine birds. *Conservation Biology* 9:1059–71.

Knopf, F. L. 1996a. Mountain plover (*Charadrius montanus*). In *The Birds of North America*, no. 211, ed. A. Poole and F. Gill. Philadelphia: Birds of North America; Washington DC: American Ornithologists' Union.

Knopf, F. L 1996b. Prairie legacies – birds. In *Prairie Conservation: Conserving North America's Most Endangered Ecosystem*, ed. F. B. Samson and F. L. Knopf, 13–48. Covelo CA: Island Press.

————, and J. R. Rupert. 1996. Reproduction and movements of mountain plovers breeding in Colorado. *Wilson Bulletin* 108:504–6.

Knowles, C. J., and P. R. Knowles. 1984. Additional records of mountain plovers using prairie dog towns in Montana. *Prairie Naturalist* 16(4): 183–86.

Knowles, C. J., C. J. Stoner, and S. P. Gieb. 1982. Selective use of black-tailed prairie dog towns by mountain plovers. *Condor* 84(1): 71–74.

Kotliar, N. B. 2000. Application of the new keystone-species concept to prairie dogs: How well does it work? *Conservation Biology* 14:1715–21.

————, B. W. Baker, A. D. Whicker, and G. Plumb. 1999. A critical review of assumptions about the prairie dog as a keystone species. *Environmental Management* 24:177–92.

Krueger, K. A. 1984. An experimental analysis of interspecific feeding relationships among bison, pronghorn and prairie dog. Abstract. *Bulletin of the Ecological Society of America* 65:267.

————. 1986. Feeding relationships among bison, pronghorn and prairie dogs: An experimental analysis. *Ecology* 67:760–70.

Loether, P. E., H. D. Douglas III, and C. L. Gratto-Trevor. 2001. Willet (*Catoptrophorus semipalmatus*). In *The Birds of North America*, no. 570, ed. A. Poole and F. Gill. Philadelphia: Birds of North America; Washington DC: American Ornithologists' Union.

Manzano-Fischer, P., R. List, and G. Ceballos. 1999. Grassland birds in prairie-dog towns in northwestern Chihuahua, Mexico. *Studies in Avian Biology* 19:263–71. In *Ecology and Conservation of Grassland Birds of the Western Hemisphere*, ed. P. D. Vickery and J. R. Herkert. Camarillo CA: Cooper Ornithological Society.

McCaffery. B. J., T. A. Sordahl, and P. Zahler. 1994. Behavioral ecology of the mountain plover in northeastern Colorado. *Wader Study Group Bulletin* 40:18–21.

Miller, B., G. Ceballos, and R. P. Reading. 1994. The prairie dog and biotic diversity. *Conservation Biology* 8:677–81.

———, R. Reading, J. Hoogland, T. Clark, G. Ceballos, R. List, S. Forrest, L. Hanebury, P. Manzano, J. Pacheco, and D. Uresk. 2000. The role of prairie dogs as a keystone species: Response to Stapp. *Conservation Biology* 14:318–21.

Moriarty, L. J. 1965. A study of the breeding biology of the chestnut-collared longspur (*Calcarius ornatus*) in northeastern South Dakota. *South Dakota Bird Notes* 17:76–79.

Nowicki, T. 1973. A behavioral study of the marbled godwit in North Dakota. Master's thesis, Central Michigan University.

Olson, S. L. 1985. Mountain plover food items on and adjacent to a prairie dog town. *Prairie Naturalist* 17:83–90.

Olson-Edge, S. L., and W. D. Edge. 1987. Density and distribution of the mountain plover on the Charles M. Russell National Wildlife Refuge. *Prairie Naturalist* 19:233–38.

Pampush, G. J, and R. G. Anthony. 1993. Nest success, habitat utilization, and nest-site selection of long-billed curlews in the Columbia Basin, Oregon. *Condor* 95:957–67.

Parrish, T. L., S. H. Anderson, and W. F. Oelklaus. 1993. Mountain plover habitat selection in the Powder River Basin, Wyoming. *Prairie Naturalist* 25:219–26.

Peterjohn, B. G., and J. R. Sauer. 1999. Population status of North American grassland birds from the North American Breeding Bird Survey, 1966–1996. *Studies in Avian Biology* 19:27–44. In *Ecology and Conservation of Grassland Birds of the Western Hemisphere*, ed. P. D. Vickery and J. R. Herkert. Camarillo CA: Cooper Ornithological Society.

Petraitis, P. S., R. E. Latham, and R. A. Niesenbaum. 1989. The maintenance of species diversity by disturbance. *Quarterly Review of Biology* 64:292–419.

Redmond, R. L, and D. A. Jenni. 1986. Population ecology of the long-billed curlew (*Numenius americanus*) in western Idaho. *Auk* 103:755–67.

Robbins, M. B. 1999. Sprague's pipit (*Anthus spragueii*). In *The Birds of North America*, no. 439, ed. A. Poole and F. Gill. Philadelphia: Birds of North America; Washington DC: American Ornithologists' Union.

Ryan, M. R., R. B. Renken, and J. J. Dinsmore. 1984. Marbled godwit habitat selection in the northern prairie region. *Journal of Wildlife Management* 48:1206–18.

Severe, D. S. 1977. Revegetation of black-tailed prairie dog mounds on shortgrass prairie in Colorado. Master's thesis, Colorado State University.

Shackford, J. S. 1991. Breeding ecology of the mountain plover in Oklahoma. *Bulletin of the Oklahoma Ornithological Society* 24:9–13.

Shackford, J. S., and J. D. Tyler. 1991. *Vertebrates Associated with Black-tailed Prairie Dog Colonies in Oklahoma*. Oklahoma City: Non-game Program, Oklahoma Department of Wildlife Conservation.

Shane, T. G. 2000. Lark bunting (*Calamospiza melanocorys*). In *The Birds of North America*, no. 543, ed. A. Poole and F. Gill. Philadelphia: Birds of North America; Washington DC: American Ornithologists' Union.

Sharps, J. C., and D. W. Uresk. 1990. Ecological review of black-tailed prairie dogs and associated species in western South Dakota. *Great Basin Naturalist* 50:339–45.

Snell, G. P. 1985. Results of control of prairie dogs. *Rangelands* 7:30.

Stapp, P. 1998. A reevaluation of the role of prairie dogs in Great Plains grasslands. *Conservation Biology* 12:1253–59.

Strong, M. A. 1971. Avian productivity on the shortgrass prairie of north-central Colorado. Master's thesis, Colorado State University.

Taylor, S. V., and V. M. Ashe. 1976. The flight display and other behavior of male lark buntings (*Calamospiza melanocorys*). *Bulletin of the Psychonomic Society* 7:527–29.

Thompson, L. S., and D. Sullivan. 1979. Breeding birds of prairie grassland and shrubland habitats in northeastern Montana – 1978. *American Birds* 33:88–89.

Tyler, J. D. 1968. Distribution and vertebrate associates of the black-tailed prairie dog in Oklahoma. PhD diss., University of Oklahoma.

———. 1970. Vertebrates in a prairie dog town. *Proceedings of the Oklahoma Academy of Science* 50:110–13.

———. Additional vertebrates in a Jackson County prairie dog town. Unpublished manuscript.

Uresk, D. W., G. L. Shenbeck, and J. T. O'Rouke, eds. 1997. *Conserving Biodiversity on Native Rangelands: Symposium Proceedings*. General Technical Report RM-GTR-298. Fort Collins CO: U.S. Forest Service.

Vickery, P. D., P. L. Tubaro, J. M. Cardoso de Silva, B. P. Peterjohn, J. R. Herkert, and R. B. Cavalcanti. 1999. Conservation of grassland birds in the Western Hemisphere. *Studies in Avian Biology* 19:2–26. In *Ecology and Conservation of Grassland Birds of the Western Hemisphere*, ed. P. D. Vickery and J. R. Herkert. Camarillo CA: Cooper Ornithological Society.

Wilcomb, M. J. 1954. A study of prairie dog burrow systems and the ecology of their arthropod inhabitants in central Oklahoma. PhD diss., University of Oklahoma.

With, K. A. 1994. McCown's longspur (*Calcarius mccownii*). In *The Birds of North America*, no. 96, ed. A. Poole and F. Gill. Philadelphia: Birds of North America; Washington DC: American Ornithologists' Union.

Wormer, J. V. 1969. *The World of the Pronghorn*. Philadelphia: Lippincott.

7. THE HIGH PLAINS RAPTORS

For a general literature review of raptors, see Johnsgard (1990). Individual species surveys exist for the prairie falcon (Steenhof 1998, Anderson and Squires 1997), ferruginous hawk (Bechard and Schmutz 1995, Dechant et al. 1999, 2000a, 2000b), northern harrier (McWriter and Bildstein

1996, Dechant et al. 1999, 2000a, 2000b), Swainson's hawk (England, Bechard, and Houston 1997, Dechant et al. 1999, 2000a, 2000b), and golden eagle (Phillips et al. 1990, Kochert et al. 2002).

Allen, G. T. 1987a. Estimating prairie falcon and golden eagle nesting populations in North Dakota. *Journal of Wildlife Management* 51:739–44.

———. 1987b. Prairie falcon aerie site characteristics and aerie use in North Dakota. *Condor* 89:187–90.

Allison, P. S., A. W. Leary, and M. J. Bechard. 1995. Observations of wintering ferruginous hawks (*Buteo regalis*) feeding on prairie dogs (*Cynomys ludovicianus*) in the Texas panhandle. *Texas Journal of Science* 47:235–37.

Anderson, S. H., and J. R. Squires. 1997. *The Prairie Falcon.* Austin: University of Texas Press.

Bailey, F. M. 1928. *Birds of New Mexico.* Santa Fe: New Mexico Department of Fish & Game.

Becker, D. M. 1979. A survey of raptors on national forest land in Carter County, Montana. Final report (unpublished), U.S. Forest Service, Northern Region.

Bechard, M. J., and J. K. Schmutz. 1995. Ferruginous hawk (*Buteo regalis*). In *The Birds of North America*, no. 172, ed. A. Poole and F. Gill. Philadelphia: Birds of North America; Washington DC: American Ornithologists' Union.

Blair, C. L., and F. Schitoskey Jr. 1982. Breeding biology and diet of the ferruginous hawk in South Dakota. *Wilson Bulletin* 94:46–54.

Brown, J. E. 1992. *Animals of the Soul: Sacred Animals of the Oglala Sioux.* Rockport MA: Element.

Cully, J. F., Jr. 1991. Response of raptors to reduction of Gunnison's prairie dog population by plague. *American Midland Naturalist* 125:140–49.

Dechant, D. J., M. L. Sondreal, D. H. Johnson, L. D. Igl, C. M. Goldade, M. P. Nenneman, and B. R. Euliss. 1999 (rev. 2001). *Effects of Management Practices on Grassland Birds: Ferruginous Hawk.* Jamestown ND: Northern Prairie Wildlife Research Center.

———. 2000a (rev. 2001). *Effects of Management Practices on Grassland Birds: Northern Harrier.* Jamestown ND: Northern Prairie Wildlife Research Center.

———. 2000b (rev. 2001). *Effects of Management Practices on Grassland Birds: Swainson's Hawk.* Jamestown ND: Northern Prairie Wildlife Research Center.

Edwards, B. E. 1973. A nesting study of a small population of prairie falcons in southern Alberta. *Canadian Field-Naturalist* 87:322–24.

England, A. S., M. J. Bechard, and C. S. Houston. 1997. Swainson's hawk (*Buteo swainsoni*). In *The Birds of North America*, no. 265, ed. A. Poole and F. Gill. Philadelphia: Birds of North America; Washington DC: American Ornithologists' Union.

Ensign, J. T. 1983. Nest site selection, productivity and food habits of ferruginous hawks in southeastern Montana. Master's thesis, Montana State University.

Gilmer, D. S., and R. E. Stewart. 1983. Ferruginous hawk populations and habitat use in North Dakota. *Journal of Wildlife Management* 47:146–57.

Glinski, R. L., ed. 1998. *The Raptors of Arizona*. Tucson: University of Arizona Press.

Houston, S. C., D. G. Smith, and C. Rohner. 1998. Great horned owl (*Bubo virginianus*). In *The Birds of North America*, no. 372, ed. A. Poole and F. Gill. Philadelphia: Birds of North America; Washington DC: American Ornithologists' Union.

Johnsgard, P. A. 1990. *Hawks, Eagles and Falcons of North America*. Washington DC: Smithsonian Institution Press.

Kochert, M. N., K. Steenhof, C. L. McIntyre, and E. H. Craig. 2002. Golden eagle (*Aquila chrysaetos*). In *The Birds of North America*, no. 684, ed. A. Poole and F. Gill. Philadelphia: Birds of North America; Washington DC: American Ornithologists' Union.

Lokemoen, J. T., and H. F. Duebbert. 1976. Ferruginous hawk nesting ecology and raptor populations in northern South Dakota. *Condor* 78:464–70.

Luttich, S. N., L. B. Keith, and J. D. Stephenson. 1971. Population dynamics of the red-tailed hawk (*Buteo jamaicensis*) at Rochester, Alberta. *Auk* 88:75–87.

Manci, K. M. 1992. Winter raptor use of urban prairie dog colonies. Abstract. *Journal of the Colorado Field Ornithologists* 26:132.

McLaren, P. A., S. H. Anderson, and D. E. Runde. 1988. Food habits and nest characteristics of breeding raptors in southwestern Wyoming. *Great Basin Naturalist* 48:548–53.

McWriter, R. B., and K. L. Bildstein. 1996. Northern harrier (*Circus hudsonicus*). In *The Birds of North America*, no. 210, ed. A. Poole and F. Gill. Philadelphia: Birds of North America; Washington DC: American Ornithologists' Union.

Olendorff, R. R. 1972. *The Large Birds of Prey of the Pawnee National Grasslands: Nesting Habits and Productivity, 1969–71*. Grassland Biome Technical Report 151. Fort Collins CO: U.S. International Biological Program.

———. 1993. *Status, Biology and Management of Ferruginous Hawks: A Review*. Special Report, U.S. Bureau of Land Management. Boise ID: Raptor Research and Technical Assistance Center.

Phillips, R. L., A. H. Wheeler, J. M. Lockhart, T. P. McEnerey, and N. C. Forrester. 1990. *Nesting Ecology of Golden Eagles and Other Raptors in Southeastern Montana and Northern Wyoming*. Technical Bulletin 26. Washington DC: U.S. Fish & Wildlife Service.

Plumpton, D. L., and D. E. Anderson. 1998. Anthropogenic effects on winter behavior of ferruginous hawks. *Journal of Wildlife Management* 62:340–46.

Roth, S. D., Jr., and J. M. Marzluff. 1989. Nest placement and productivity of ferruginous hawks in Kansas. *Transactions of the Kansas Academy of Science* 92:132–48.

Runde, D. E., and S. A. Anderson. 1986. Characteristics of cliff and nest sites used by breeding prairie falcons. *Journal of Raptor Research* 20:21–28.

Seery, D. J., and D. J. Matiatos. 2000. Response of wintering buteos to plague epizootic in prairie dogs. *Western North American Naturalist* 4:420–25.

Squires, J. R., S. A. Anderson, and R. Oakleaf. 1989. Food habits of prairie falcons in Campbell County, Wyoming. *Journal of Raptor Research* 23:157–61.

———. 1993. Home range size and habitat use of nesting prairie falcons near oil developments in northeastern Wyoming. *Journal of Field Ornithology* 64:1–10.

Steenhof, K. 1998. Prairie falcon (*Falco mexicanus*). In *The Birds of North America*, no. 348, ed. A. Poole and F. Gill. Philadelphia: Birds of North America; Washington DC: American Ornithologists' Union.

Watson, J. 1997. *The Golden Eagle*. Princeton NJ: Princeton University Press.

Webster, H. M., Jr. 1944. A survey of the prairie falcon in Colorado. *Auk* 61:609–16.

Zelenak, J. R., and J. J. Rotella, 1997. Nest success and productivity of ferruginous hawks in northern Montana. *Canadian Journal of Zoology* 75:1035–41.

8. THE VARMINT AND PREDATOR WARS

Adams, C. C. 1930. Rational predatory animal control. *Journal of Mammalogy* 11:353–62.

Arthur, L. M., R. L. Gunn, E. H. Carpenter, and W. W. Shaw. 1977. Predator control: The public view. *Transactions of the North American Wildlife and Natural Resources Conference* 42:137–45.

Bell, W. R. 1919. Cooperative campaigns for the control of ground squirrels, prairie dogs and jackrabbits. *U.S. Dept. of Agriculture Yearbook, 1917,* 225–33.

———. 1921. Death to the rodents. *U.S. Dept. of Agriculture Yearbook, 1920,* 421–28.

Cain, S. A. 1978. Predator and pest control. In *Wildlife in America: Contributions to an Understanding of American Wildlife and Its Conservation*, ed. H. P. Brokaw, 379–95. Washington DC: U.S. Fish & Wildlife Service, U.S. Forest Service, and National Oceanic and Atmospheric Administration.

Clawson, M. 1971. *The Bureau of Land Management*. New York: Praeger.

Collins, A. R., J. P. Workman, and D. W. Uresk. 1984. An economic analysis of black-tailed prairie dog (*Cynomys ludovicianus*) control. *Journal of Range Management* 37:358–61.

Colorado Division of Wildlife. 2003. *Conservation Plan for Grassland Species in Colorado.* Denver: Colorado Division of Wildlife.

Deisch, M. S., D. W. Uresk, and R. L. Linder. 1989. Effects of two prairie dog rodenticides on ground-dwelling invertebrates in western South Dakota. In *Proceedings Ninth Great Plains Wildlife Damage Control Workshop* (April 17–20, 1988, Fort Collins CO), ed. D. W. Uresk et al., 166–70. General Technical Report RM-171. Fort Collins CO: U.S. Forest Service.

DiSilvestro, R. L. 1985. The federal animal damage control program. In *Audubon Wildlife Report, 1985*, ed. R. L. DiSilvestro, 130–48. New York: National Audubon Society.

Dobie, J. F. 1949. *The Voice of the Coyote.* New York: Little, Brown.

Fisher, H. 1982. War on the dog towns. *Defenders* 57:9–12.

Glinski, R. L., ed. 1998. *The Raptors of Arizona.* Tucson: University of Arizona Press.

Grady, W. 1994. *The Nature of Coyotes*. Vancouver: Douglas & McIntyre.

Hall, E. R. 1930. Predatory animal destruction. *Journal of Mammalogy* 11:362–77.

Henderson. 1930. The control of the coyote. *Journal of Mammalogy* 11:336–53.

Henke, S. E., and F. C. Bryant. 1999. Effects of coyote removal on the faunal community in eastern Texas. *Journal of Wildlife Management* 63:1066–81.

Hoogland, J. L. 1994. *The Black-tailed Prairie Dog: Social Life of a Burrowing Mammal.* Chicago: University of Chicago Press.

Howell, A. B. 1930. At the cross-roads. *Journal of Mammalogy* 11:377–89.

Hyngstrom, S. E., and P. M. MacDonald. 1989. Efficacy of three formulations of zinc phosphide for black-tailed prairie dog control. Abstract. In *Proceedings Ninth Great Plains Wildlife Damage Control Workshop* (April 17–20, 1988, Fort Collins CO), ed. D. W. Uresk et al., 181. General Technical Report RM-171. Fort Collins CO: U.S. Forest Service.

Kayser, M. 1998. Have varmint rifle will travel. *American Hunter* (June), 44–62.

Knowles, C. J. 1988. An evaluation of shooting and habitat alteration for control of black-tailed prairie dogs. In *Proceedings Eighth Great Plains Wildlife Damage Control Workshop* (April 28–30, 1987, Rapid City SD), ed. D. W. Uresk et al., 53–56. General Technical Report RM-154. Fort Collins CO: U.S. Forest Service.

———. 2002. *Status of White-tailed and Gunnison's Prairie Dogs*. Missoula MT: National Wildlife Federation; Washington DC: Environmental Defense.

Krueger, K. 1988. Prairie dog overpopulation: Value judgment or ecological reality? In *Proceedings Eighth Great Plains Wildlife Damage Control Workshop* (April 28–30, 1987, Rapid City SD), ed. D. W. Uresk et al., 39–45. General Technical Report RM-154. Fort Collins CO: U.S. Forest Service.

Leopold, A. S., S. A. Cain, C. Cottam, I. N. Gabrielson, and T. L. Kimball. 1963. Predator and rodent control in the United States. *Transactions of the North American Wildlife Conference* 29:27–49.

Luce, R., ed. 2003. *A Multi-state Conservation Plan for the Black-tailed Prairie Dog*, Cynomys ludovicianus, *in the United States – an Addendum to the Black-tailed Prairie Dog Conservation Assessment and Strategy, November 3, 1999*. Washington DC: Wildlife Management Institute (WMI).

Marsh, R. E. 1989. Relevant characteristics of zinc phosphide as a rodenticide. In *Proceedings Ninth Great Plains Wildlife Damage Control Workshop* (April 17–20, 1988, Fort Collins CO), ed. D. W. Uresk et al., 70–74. General Technical Report RM-171. Fort Collins CO: U.S. Forest Service.

Merriam, C. H. 1902. The prairie dog of the Great Plains. In *Yearbook of the United States Department of Agriculture, 1901*, 257–70. Washington DC: U.S. Department of Agriculture.

Merriam, C. H. 1902. The prairie dog of the Great Plains. In *Yearbook of the United*

States Department of Agriculture, 1901, 257–70. Washington DC: U.S. Department of Agriculture.

McNulty, F. 1971. *Must They Die? The Strange Case of the Prairie Dog and the Black-footed Ferret*. Garden City NY: Doubleday.

Moline, P. R., and S. Demarais. 1988. Efficacy of aluminum phosphide for black-tailed prairie dog and yellow-faced pocket gopher control. In *Proceedings Eighth Great Plains Wildlife Damage Control Workshop* (April 28–30, 1987, Rapid City SD), ed. D. W. Uresk et al., pp. 66–67. General Technical Report RM-154. Fort Collins CO: U.S. Forest Service.

Murie, A. 1940. *Ecology of the Coyote of the Yellowstone*. Fauna Series No. 4. Washington DC: U.S. Government Printing Office.

Nelson, E. W. 1919. Annual report to Chief of Biological Survey. In *U.S. Dept. Agriculture Annual; Rept. for 1919*, 275–98. Washington DC: U.S. Department of Agriculture.

Oakes, C. L. 2000. History and consequences of keystone mammal eradication in the desert grasslands: The Arizona black-tailed prairie dog (*Cynomys ludovicianus arizonicus*). PhD diss., University of Texas at Austin.

Olson, J. 1971. *Slaughter the Animals, Poison the Earth*. New York: Simon & Schuster.

Randall, D. 1976. Shoot the damn prairie dogs. *Defenders* 51:378–81.

———. 1976. Poison the damn prairie dogs. *Defenders* 51:381–83.

Raventon, E. 1994. *Island in the Plains: A Black Hills Natural History*. Boulder CO: Johnson Printing.

Roemer, D. M., and S. C. Forrest. 1996. Prairie dog poisoning in northern Great Plains: An analysis of programs and policies. *Environmental Management* 20:349–59.

Rosmarino, N. J. 2003. Comments re annual black-tailed prairie dog (*Cynomys ludovicianus*) status review information request. Letter of February 3, 2003, to U.S. Fish & Wildlife Service, Pierre SD.

———, E. Robertson, and D. Crawford. 2003. Comments re draft Grassland Species Conservation Plan. Letter of October 13, 2003, to Colorado Division of Wildlife, for Forest Guardians, Center for Native Ecosystems, and Rocky Mountain Animal Defense.

Schlebecker, J. T. 1963. *Cattle Raising on the Plains: 1900–1961*. Lincoln: University of Nebraska Press.

Ryden, H. 1979. *God's Dog: A Celebration of the North American Coyote*. New York: Coward, McCann & Geohagen.

Sharps, J. 1988. Politics, prairie dogs and the sportsman. In *Proceedings Eighth Great Plains Wildlife Damage Control Workshop* (April 28–30, 1987, Rapid City SD), ed. D. W. Uresk et al., 117–18. General Technical Report RM-154. Fort Collins CO: U.S. Forest Service.

Snell, G. P. 1985. Results of control of prairie dogs. *Rangelands* 7:30.

———, and B. D. Hlavachick. 1980. Control of prairie dogs – the easy way. *Rangelands* 2:239–40.

South Dakota Department of Agriculture. "2001 South Dakota Prairie Dog Shooting." http://www.state.sd.us/doa/prairiedog/pdshooting.htm.

Taylor, W. P., and J. V. G. Loftfield. 1924. Damage to range grasses by the Zuni prairie dog. *Bulletin of the U.S. Department of Agriculture* 1227:1–16.

Uresk, D. W. 1985. Effects of controlling black-tailed prairie dogs on plant production. *Journal of Range Management* 38:466–68.

U.S. Department of Agriculture, Wildlife Services. 2003. Annual tables of federal wildlife control activities.

Vosburgh, T. C, and L. R. Irby. 1998. Effects of recreational shooting of prairie dog colonies. *Journal of Wildlife Management* 62:363–72. (See also 62:1153.)

9. TAMING THE GREAT AMERICAN DESERT

Abbey, E. 1988. *One Life at a Time, Please.* New York: Henry Holt.

Anonymous. 2000. Why bison? *Bison World* 25(1): 46.

Callenbach, K. 2000. *Bring Back the Buffalo! A Sustainable Future for America's Great Plains.* Berkeley: University of California Press.

Clawson, M. 1971. *The Bureau of Land Management.* New York: Praeger.

Conley, K., and S. Albrecht. 2000. History of the bison industry. *Bison World* 25(1): 21–22.

Flores, D. 2003. *The Natural West: Environmental History in the Great Plains and Rocky Mountains.* Norman: University of Oklahoma Press.

Frazier, I. 1989. *Great Plains.* New York: Farrar Straus Giroux.

Küchler, A. W. 1966. *Potential Natural Vegetation of the Coterminous United States.* Special Publication 35. New York: American Geographic Society.

Lang, R. E., D. E. Popper, and F. J. Popper. 1995. Progress of the nation: The settlement history of the enduring American frontier. *Western Historical Quarterly* (Autumn 1995).

Luce, R., ed. 2003. *A Multi-state Conservation Plan for the Black-tailed Prairie Dog*, Cynomys ludovicianus, *in the United States – an Addendum to the Black-tailed Prairie Dog Conservation Assessment and Strategy, November 3, 1999.* Washington DC: Wildlife Management Institute (WMI).

Lyman, H. F., and G. Metzer. 1998. *Mad Cowboy: Plain Truth from the Cattle Rancher Who Won't Eat Meat.* New York: Scribner's.

McHugh, T. 1972. *The Time of the Buffalo.* New York: Knopf.

Risser, P. G., E. C. Birney, H. D. Bloeker, S. W. May, W. J. Parton, and J. A. Weins. 1981. *The True Prairie Ecosystem.* Stroudsburg PA: Hutchinson Russ.

Samson, F. B., and F. L. Knopf. 1994. Prairie conservation in North America. *BioScience* 44:418–21.

Schlebecker, J. T. 1963. *Cattle Raising on the Plains: 1900–1961.* Lincoln: University of Nebraska Press.

Stegner, W. 1962. *Wolf Willow: A History, a Story and a Memory of the Last Plains Frontier.* New York: Viking.

———. 1992. *The American West as Living Space.* Ann Arbor: University of Michigan Press.

Webb, W. P. 1931. *The Great Plains.* New York: Grosset & Dunlap.

Zaslowsky, D., and the Wilderness Society. 1986. *These American Lands.* New York: Henry Holt.

10. THE USFS, BLM, AND BIA

Abbey, E. 1988. *One Life at a Time, Please.* New York: Henry Holt.

Brower, K. 1997. *Our National Forests: American Legacy.* Washington DC: National Geographic Society.

Cable, K. A., and R. M. Tim. 1988. Efficacy of deferred grazing in reducing prairie dog reinfestation rates. In *Proceedings Eighth Wildlife Damage Control Workshop* (April 28–30, 1987, Rapid City SC), ed. D. W. Uresk et al., 46–49. General Technical Report RM-154. Fort Collins CO: U.S. Forest Service.

Callenbach, K. 2000. *Bring Back the Buffalo! A Sustainable Future for America's Great Plains.* Berkeley: University of California Press.

Clark, T. W. 1970. Some prairie dog–range relationships in the Laramie Plains of Wyoming. *Journal of Wyoming Range Management* 282:40–51.

Clawson, M. 1971. *The Bureau of Land Management.* New York: Praeger.

Coppock, D. L., J. E. Ellis, J. K. Detling, and M. I. Dyer. 1983. Plant-herbivore interactions in a North American mixed-grass prairie. *Oecologia* (Berlin) 56:1–15.

Finch, D. M. 1992. *Threatened, Endangered and Vulnerable Species of Terrestrial Vertebrates in the Rocky Mountain Region.* General Technical Report RM-215. Fort Collins CO: U.S. Forest Service.

Guenther, D. A. 2000. Cattle use of prairie dog towns on the shortgrass plains of Colorado. Master's thesis, Colorado State University.

Hansen, R. M., and I. K. Gold. 1977. Blacktail prairie dogs, desert cottontails and cattle trophic relations on shortgrass range. *Journal of Range Management* 30:210–14.

Howard, W. E., K. A. Wagnon, and J. R. Bentley Jr. 1959. Competition between ground squirrels and cattle for range forage. *Journal of Range Management* 12:110–23.

Hyde, R. M. 1981. Prairie dogs and their influence on rangeland and livestock. In *Proceedings Fifth Great Plains Damage Control Workshop*, ed. R. M. Timm and R. J. Johnson, 202–6. Lincoln: Institute of Agriculture and Natural Resources, University of Nebraska.

Jacobs, L. 1991. *Waste of the West: Public Lands Ranching.* Tucson: Free Our Public Lands.

Johnsgard, P. A. 2002. *Grassland Grouse and Their Conservation.* Washington DC: Smithsonian Institution Press.

Knowles, C. J. 1986. Some relationships of black-tailed prairie dogs to livestock grazing. *Great Basin Naturalist* 46:198–203.

Koford, C. B. 1958. Prairie dogs, whitefaces and blue grama. *Wildlife Monographs* 3:1–78.

Lerwick, A. C. 1974. The effects of the black-tailed prairie dog on vegetative composition and their diet in relation to cattle. Master's thesis, Colorado State University.

Linderman, F. B. 1930. *American: The Life Story of a Great Indian, Plenty-coups, Chief of the Crows.* New York: John Day.

Lyman, H. F., and G. Metzer. 1998. *Mad Cowboy: Plain Truth from the Cattle Rancher Who Won't Eat Meat.* New York: Scribner's.

Manning, R. 1991. *Last Stand.* New York: Penguin Books.

Milchunas, D. G., W. K. Lauenroth, and I. C. Burke. 1998. Livestock grazing: Animal and plant diversity of shortgrass steppe and the relationship to ecosystem function. *Oikos* 83:65–74.

O'Melia, M. E. 1980. Competition between prairie dogs and beef cattle for range forage. Master's thesis, Oklahoma State University.

———, F. Knopf, and J. Lewis. 1982. Some consequences of competition between prairie dogs and beef cattle. *Journal of Range Management* 35:580–85.

Osgood, E. S. 1929. *The Day of the Cattlemen.* Minneapolis: University of Minnesota Press.

Phillips, P. G. 1936. The distribution of rodents in overgrazed and normal grasslands in central Oklahoma. *Ecology* 17:673–79.

Predator Conservation Alliance. 2003. *Restoring the Prairie Dog Ecosystem of the Great Plains.* Bozeman MT: Predator Conservation Alliance.

Pritzker, B. M. 1998. *Native Americans: An Encyclopedia of History, Culture and Peoples.* 2 vols. Santa Barbara CA: ABC-CLIO.

Prucha, F. B. 1990. *Atlas of American Indian Affairs.* Lincoln: University of Nebraska Press.

Rifkin, J. 1992. *Beyond Beef: The Rise and Fall of the Cattle Culture.* New York: Plume.

Roehrs, Z. P. 2004. Biogeography and population dynamics of the prairie dog *Cynomys ludovicianus* Ord in Nebraska from 1965 to 2003. Master's thesis, University of Nebraska–Lincoln.

Rowley, W. D. 1985. *U.S. Forest Service Grazing and Rangelands.* College Station: Texas A&M University Press.

Ryden, H. 1977. *God's Dog: A Celebration of the North American Coyote.* New York: Coward, McCann & Geohagen.

Samson, F. B., and W. R. Ostlie. 1998. Grasslands. In *Status and Trends of the Nation's Biological Resources*, ed. M. J. Mace et al, 437–71. 2 vols. Reston VA: U.S. Department of the Interior, U.S. Geological Survey.

Schlebecker, J. T. 1963. *Cattle Raising on the Plains: 1900–1961.* Lincoln: University of Nebraska Press.

Secretary of Agriculture. 1936. *The Western Range.* Washington DC: U.S. Government Printing Office.

Stahl, A. 2003. Rules, rules, rules, rules. *Inner Voice: Newsletter of Forest Service Employees and Environmental Ethics* 5(2). Reprinted in *Forest Magazine* (spring 2003): 29–31.

Stegner, W. 1992. *The American West as Living Space.* Ann Arbor: University of Michigan Press.

Uresk, D. W. 1984. Black-tailed prairie dog food habits and forage relationships in western South Dakota. *Journal of Range Management* 37:325–29.

U.S. Department of Agriculture. 1949. *Trees: Yearbook of Agriculture, 1949.* Washington DC: U.S. Department of Agriculture.

Utley, R. M., and W. E. Washburn. 1977. *The American Heritage History of the Indian Wars.* New York: American Heritage.

Waldman, C. 1985. *Atlas of the North American Indian.* New York: Facts on File.

Walker, J. R. 1917. The Sun Dance and other ceremonies of the Oglala Division of the Dakota Teton. *Anthropological Papers of the American Museum of Natural History* 16(2).

Zaslowsky, D., and the Wilderness Society. 1986. *These American Lands.* New York: Henry Holt.

11. THE GREAT PLAINS GRASSLAND ECOSYSTEM

Adelman, C., and B. L. Schwartz. 2001. *Prairie Directory of North America.* Wilmette IL: Lawndale Enterprises.

American Society of Mammalogists. 1998. Resolution on the decline of the prairie dogs and the grassland ecosystem of North America. *Journal of Mammalogy* 29:1447–48.

Bachard, R. R. 2001. *The American Prairie: Going, Going, Gone?* Boulder CO: National Wildlife Federation.

Burroughs, R. D. 1961. *The Natural History of the Lewis and Clark Expedition.* East Lansing: Michigan State University Press.

Colorado Division of Wildlife. 2003. *Conservation Plan for Grassland Species in Colorado.* Denver: Colorado Division of Wildlife.

Coupland, R. T. 1950. Ecology of the mixed prairie in Canada. *Ecological Monographs* 20:272–315.

———, and T. C. Brayshaw. 1953. The fescue grassland in Saskatchewan. *Ecology* 34:386–405.

Crump, D. J., ed. 1984. *A Guide to Our Federal Lands.* Washington DC: National Geographic Society.

Cushman, R. C., and S. R. Jones. 1988. *The Shortgrass Prairie.* Boulder CO: Pruett.

Dolby, C. C. 1999. The national grasslands and disappearing biodiversity: Can the prairie dog save us from an ecological desert? *Environmental Law* 29:213–34.

Donahue, D. L. 2000. *The Western Range Revisited: Removing Livestock from Public Lands to Conserve Native Biodiversity.* Norman: University of Oklahoma Press.

Ferguson, D., and N. Ferguson. 1983. *Sacred Cows at the Public Trough.* Bend OR: Maverick.

Finch, D. M. 1992. *Threatened, Endangered and Vulnerable Species of Terrestrial Vertebrates in the Rocky Mountain Region.* General Technical Report RM-215. Fort Collins CO: U.S. Forest Service.

Foss, P. O. 1969. *Politics and Grass: The Administration of Grazing on the Public Domain.* New York: Greenwood.

Jacobs, L. 1991. *Waste of the West: Public Lands Ranching.* Tucson: Free Our Public Lands.

Jones, S. R., and R. C. Cushman. 2004. *A Field Guide to the North American Prairie.* Boston: Houghton Mifflin.

Johnsgard, P. A. 2003. *Great Wildlife of the Great Plains.* Lawrence: University Press of Kansas.

———. 2003. *Lewis and Clark on the Great Plains: A Natural History.* Lincoln: University of Nebraska Press.

Knopf, F. L., and F. B. Samson, eds. 1997. *Ecology and Conservation of Great Plains Vertebrates.* New York: Springer.

Küchler, A. W. 1966. *Potential Natural Vegetation of the Coterminous United States.* Special Publication 35. New York: American Geographic Society.

LaRoe, E. T., G. S. Farris, C. E. Puckett, P. D. Doran, and M. J. Mac, eds. 1995. *Our Living Resources: A Report to the Nation on the Distribution, Abundance and Health of U.S. Plants, Animals and Ecosystems.* Washington DC: National Biological Service.

Manning. R. 1995. *Grassland: The History, Biology, Politics and Promise of the American Prairie.* New York: Penguin Books.

Ostlie, W. R., R. E. Schneider, J. M. Aldrich, T. M. Faust, R. L. B. McKim, and S. I. Chaplin. 1997. *The Status of Biodiversity in the Great Plains.* Arlington VA: The Nature Conservancy.

Popper, D. E., and F. J. Popper. 1987. The Great Plains: From dust to dust. *Planning* 53 (December): 12–17.

Prucha, F. B. 1990. *Atlas of American Indian Affairs.* Lincoln: University of Nebraska Press.

Reading, R. P., T. W. Clark, L. McCain, and B. J. Miller. 2002. Black-tailed prairie dog conservation: A new approach for a 21st century challenge. *Endangered Species Update* 19(4): 162–70.

Samson, F. B., and F. L. Knopf. 1994. Prairie conservation in North America. *BioScience* 44:418–21.

———, and W. R. Ostlie. 1998. Grasslands. In *Status and Trends of the Nation's Biological Resources*, ed. M. J. Mac et al., 437–71. 2 vols. Reston VA: U.S. Department of the Interior, U.S. Geological Survey.

Sieg, C. H., C. H. Flather, and S. McCanny. 1999. Recent biodiversity patterns in the Great Plains: Implications for restoration and management. *Great Plains Research* 9:277–313.

Uresk, D. W., G. L. Shenbeck, and J. T. O'Rouke, eds. 1997. *Conserving Biodiversity on Native Rangelands: Symposium Proceedings.* General Technical Report RM-GTR-298. Fort Collins CO: U.S. Forest Service.

Utley, R. M., and W. E. Washburn. 1977. *The American Heritage History of the Indian Wars.* New York: American Heritage.

APPENDIX 1. A GUIDE TO GRASSLANDS, RESERVATIONS, AND PRESERVES

Adelman, C., and B. L. Schwartz. 2001. *Prairie Directory of North America.* Wilmette IL: Lawndale Enterprises.

Cable, T. T., S. Seltman, and K. J. Cook. 1996. *Birds of Cimarron National Grassland.* General Technical Report RM-GTR-281. Fort Collins CO: U.S. Forest Service.

Hoogland, J. L. 1994. *The Black-tailed Prairie Dog: Social Life of a Burrowing Mammal.* Chicago: University of Chicago Press.

Johnsgard, P. A. 2001. *Prairie Birds: Fragile Splendor in the Great Plains.* Lawrence: University Press of Kansas.

King, J. A. 1955. Social behavior, social organization, and population dynamics in a black-tailed prairie-dog town in the Black Hills of South Dakota. In *Contributions from the Laboratory of Vertebrate Biology 67.* Ann Arbor: University of Michigan.

Winckler, S. 2004. *Prairie: A North American Guide.* Iowa City: University of Iowa Press.

Index